FINDING OUR PLACE IN THE UNIVERSE

FINDING OUR PLACE IN THE UNIVERSE

HOW WE DISCOVERED
LANIAKEA—THE MILKY WAY'S HOME

Hélène Courtois
translated by Nikki Kopelman

The MIT Press
Cambridge, Massachusetts
London, England

This book was published in French under the title *Voyage sur les flots de galaxies. Laniakea, notre nouvelle adresse dans l'Univers*, second edition, by Hélène Courtois, © 2018 Dunod, Malakoff, France. Illustrations by Rachid Maraï and Daniel Pomarède.

This book was set in Stone Serif by Westchester Publishing Services. Printed and bound in the United States of America.

Library of Congress Cataloging-in-Publication Data

Names: Courtois, Hélène (Hélène Di Nella), author.
Title: Finding our place in the universe : how we discovered Laniakea, the Milky Way's home / Hélène Courtois.
Description: Cambridge, MA : The MIT Press, [2019] | Includes bibliographical references and index.
Identifiers: LCCN 2018043980 | ISBN 9780262039956 (hardcover : alk. paper)
Subjects: LCSH: Milky Way. | Superclusters. | Galaxies. | Celestial mechanics.
Classification: LCC QB857.7 .C68 2019 | DDC 523.1/12--dc23
LC record available at https://lccn.loc.gov/2018043980

10   9   8   7   6   5   4   3   2   1

Contents

Foreword

What's a cosmographer? A cosmographer draws maps of the universe. That's how Hélène Courtois describes her work: she makes diagrams of space. There's a further temporal dimension, since, in the field of astronomy, traveling great distances also means traveling in time. Fasten your seatbelts, then. This book will take us from our immediate surroundings to destinations 500 million light-years away.

The hundreds of thousands of galaxies all around us are not distributed at regular intervals. Instead, they form dense clusters separated by vast, empty expanses and joined by galactic filaments, like an immense spiderweb. A map represents this cosmic network in two dimensions. So how is a third dimension—that is, a measure of depth—obtained?

The narrative proceeds briskly as we follow the author on her travels across the globe to work with different kinds of technology, first optical and then radio telescopes. Her research program has required collaboration with astronomers all over the world. Surfing the cosmic waves, Hélène Courtois's team watches day and night, and in every time zone. In 2009, for instance, she didn't sleep much at all because she was busy conducting 480 nighttime observations.

Devising a three-dimensional map of our Local Universe plays a key role in this process. Extremely sophisticated software is required in order to account for various aspects of observation, as well as the lack of data for certain regions. If galaxy velocities are unavailable, they can be reconstructed—like a mosaic in Pompeii missing a few tiles. Algorithms based on models and simulations of dark matter, calculation of optimal probabilities, and so-called Wiener filters that minimize "noise" enable researchers to figure out data they don't have.

All this intensive labor has yielded a map of our Local Universe and its watersheds: the Laniakea Supercluster. Remarkably, we occupy a position at the edge of the cluster bordering the Great Void. Does this herald the discovery of the Great Attractor? The reader will find out in due time.

Our author's twenty-year quest forms a tapestry with many amusing excursions. Asides on the everyday life of researchers and astronomers add a welcome, human element. The course of true research never did run smooth! Mistakes may be made along the way, but, if one perseveres, they lead to new insights. More and more, cosmological observation is a matter of enlisting vast networks of colleagues and collaborators. Our author's career illustrates this shift in the field. In the early stages, she worked with groups of fewer than ten people. Today, she is participating in teamwork on a massive scale, including preparations for the Euclid Mission—a project that involves twelve hundred people. The work at hand pays tribute to joint efforts made across national boundaries, where every field of expertise comes together to achieve a common objective.

Finally, to avoid interrupting the narrative thread and to elucidate particularly difficult points, the book features brief, standalone sidebars, which may be read separately. On every page,

Hélène Courtois draws on her pedagogical talent to explain how various distance indicators are determined, why the universe is expanding, what the terms *dark matter* and *dark energy* mean, cosmic microwave background radiation, and more. In a word, she offers the reader a brief history of the world.

Françoise Combes
Astrophysicist
French Academy of Sciences

Prologue

Our own galaxy and its neighbors are racing through the universe at the astonishing rate of several hundred kilometers per second. This fact has been known since the beginning of the 1960s; at the time, however, astrophysicists were not yet able to explain the reasons in full. Later, in the 1990s, an American research team proposed that this movement be understood in light of an enormous mass, the "Great Attractor," which, unfortunately, occupies a region that's difficult to observe.

It so happens that Lyon, where I work, is mostly celebrated for its gastronomy—but also, as you will read, for its astronomy. Searching for the Great Attractor, my team and I wound up finding the supercluster of galaxies to which our planet belongs, and in which we live. We named it Laniakea.

This book aims to share the story of this immense discovery. I hope to present a clear and simple overview of the universe and the laws of physics governing it. Accordingly, complicated mathematical notation is omitted. Using only formulas in powers of ten means sacrificing a bit of precision, but the end justifies the means: a comprehensive account, in everyday terms, of the scientific approach that guides research.

In the following pages, I will describe the methods of visualization and analysis that enable us to devise maps that, bit by bit, reveal grand-scale structures of the universe (galaxy filaments, superclusters, and voids). The reader will gain familiarity with a new, extragalactic environment (that is, spaces beyond our own galaxy), which professionals call "local" even though it extends some 500 million light-years to every side.

This book also includes the most up-to-date findings that have been made since our discovery of Laniakea, to wit, the Cosmic Velocity Web and the Dipole and Cold Spot Repellers.

The concluding portion surveys the impact this discovery has made on the field. Indeed, this program of research has promoted fuller understanding of the different ways that galaxies are formed; as such, it represents the foundation for further explorations to be conducted with multiantenna telescopes operating both on the earth and in space.

Throughout, the book seeks to recognize some of the researchers of all nationalities who have contributed, in one way or another, to the findings presented here. In particular, I have included a few portraits of exceptional women in the field of astrophysics—Henrietta Leavitt, Sandra Faber, Wendy Freedman, Vera Rubin, and Renée Kraan-Korteweg—in order to provide a different view of science and scientists. There are many more who also deserve mention. This recognition is meant to illustrate that neither place of birth nor gender matters; what counts are individual initiative and teamwork.

But these opening remarks have gone on long enough. Sit back and make yourself comfortable for a journey through space and time. The trip will follow a course of discovery that began in the Australian bush, when I was just a student, and led to my most recent adventures high above the Hawaiian palms or at observation points at the heart of a vast expanse of radio silence.

1   Our New Cosmic Address

A timeline of the scientific discoveries that have enabled cosmologists to locate galaxies in space and map the local universe in three dimensions.

You Are Here

Since 2 September 2014, it's been official: Earth has a new cosmic address. On that day, the prestigious scientific journal *Nature* published an article in which colleagues and I announced the discovery of Laniakea. This supercluster is the largest galaxy structure known to date that also includes our world. In Hawaiian, the name means "immense heaven." Indeed, its dimensions are vast and difficult to compass: it measures some 500 million light-years in diameter; in other words, light takes 500 million years to cross from one side to the other. Laniakea comprises approximately 100,000 large galaxies like our own, and about a million smaller ones, with some 100 trillion suns!

This, then, is the story of Laniakea's discovery, in which I am happy to have played an active part.

A Brief Lexicon for Understanding Cosmology

For cosmologists, the basic object of study is the galaxy. The term derives from the ancient Greek word *gala* (γάλα) for "milk." Galaxies contain stars, gas, dust, and invisible matter (also known as dark matter), all of which are held together by the effects of gravitation. Galaxies are classified by shape or size. Accordingly, there are spiral, elliptical, lenticular, irregular, dwarf, and giant galaxies. Our own galaxy, the so-called Milky Way, is relatively large: it

Figure 1.1 Examples of galaxy morphology: elliptical (ESO 325), irregular (NGC 1427A), spiral (M83).

(continued)

contains several hundred billion stars. It's a spiral galaxy, shaped like a disc with a central bulge; our sun is located at the edge of one of the spiral's branches, the "Orion Arm."

Galaxies are made up of stars. A star is "merely" a ball of gas that is extremely hot because of nuclear fusion at its core. A given star's temperature is connected to its mass: the largest stars are the

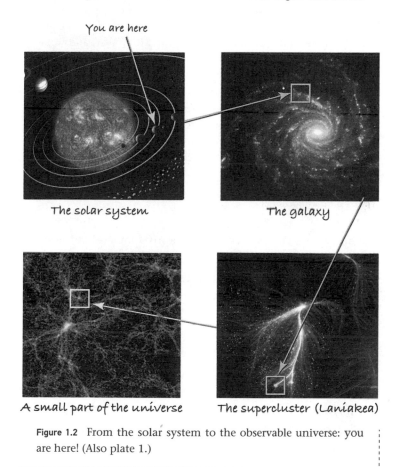

Figure 1.2 From the solar system to the observable universe: you are here! (Also plate 1.)

(continued)

(continued)

hottest, and they live for the shortest period of time. Our sun is a medium-sized star. Stars are orbited by planets: small celestial bodies that, because of their smaller dimensions, do not emit their own light. Eight planets, including Earth, orbit our sun. Some planets are orbited by even smaller bodies—such as the moon, Earth's only natural satellite.

Galaxies group together and form the universe under the effects of gravitation. We live in the Local Group, which comprises three large galaxies, including the Milky Way, and some fifty dwarf galaxies. Sometimes, galaxies group together on a larger scale to form what's known as a cluster. Our Local Group is also subject to the gravitational force of the Virgo Cluster, which contains over a thousand galaxies. Clusters occupy positions along galaxy filaments in webs that form superclusters—like Laniakea.

What's a Cosmographer?

Cosmology is a vast branch of astronomy that studies the structure and evolution of the universe since the Big Bang. To this end, cosmologists identify the celestial structures that exist in the current universe and determine how they interact. Doing so enables them to retrace the stages in which these complex bodies formed from the time the universe was very young, when matter was distributed much more evenly. Cosmologists are "historians and geographers" of the skies; however, they have a variety of specializations that differ greatly, ranging from pure theory to experimentation. My area of specialization is cosmography: creating maps of our universe. More specifically, I try to determine the position and movement of the galaxies neighboring our own—what's known as the "Local Universe" in the profession. It sounds strange to call it "local" when our closest neighbors are several hundred million light-years away! Indeed, the light we see from these galaxies was emitted when dinosaurs still walked the earth, and maybe before then. All the same, the description "local" makes sense inasmuch as even our largest maps today have managed to chart only a millionth part of the observable universe.

Whenever I'm asked to come to a school and talk about what I do, the students, whether they're younger or older, never ask *why* maps of the universe are made. What they want to know is *how* it's done. The answers to both questions are important. The answer to the first is obvious: we need a map to know where we are! After all, doesn't one need to be aware of one's current whereabouts to figure out which way to go next? Plus, knowing where you've been helps, at least a little, when trying to figure out who and what came before you. Answering the other question—*how*—proves more complicated and raises a host of other questions. What do modern-day astrophysicists do? Do they still spend their time with their eyes glued to telescopes, as Galileo did four hundred years ago? Is it still necessary to travel from one mountaintop to the next, all over the world, to gather facts for analysis and comparison with standing theories and models, so the field as a whole will advance? And what, exactly, do my duties as a university professor involve? For instance, do I have to combine my daytime teaching schedule with nighttime observations?

I begin by describing my daily routine. A key part of the job is using information technology to collect and process data (among other things). Then there's the matter of explaining the methods employed: I have to decide what zone in the celestial vault—that is, where in its concave, two-dimensional expanse—to explore with the telescope, and then gauge how far away the galaxy that interests me is, in order to bring the third dimension into the equation. Finally, a number of tricks of the trade allow me to calculate the velocity in question, making it possible to construct "dynamic maps" (as they're called) of the space that surrounds us. The students often express surprise: "I'd never have thought your job was like that!"

Basic, Day-to-Day Research

I confess I'm grateful when students ask me about how we map the universe. At the same time, I acknowledge that adults, who usually want to know *why* we do so, are raising a valid question. Their concerns are ultimately more pragmatic—after all, they're the ones who have to pay for state research! Taxpayers might think they support activities that don't produce palpable results in everyday life.

But most of what we use from day to day derives from research of one sort or another; such research may be applied or more basic in nature. Often, the discovery of some new phenomenon in the physical world prompts technology to be applied to new ends. Thus, light bulbs were invented once scientists figured out how electricity propagates and, in the process, loses energy. Some everyday items we use can also represent the results of inventing or fabricating some new tool needed for research on a fundamental level. For instance, the glass that keeps the outside of oven doors cool is directly linked to research conducted to make mirrors for very large telescopes. In order to manufacture mirrors several meters in diameter, one hundred tons of silica must be melted down; in turn, this mass (which is about a meter thick) needs to be cooled in a way that minimizes differences in temperature between the surface and the center, which cause imperfections in the glass. Efforts to ensure a high degree of thermal stability led to the invention of new kinds of glass. Results of technology developed for telescopes are now found in kitchens everywhere—and no one has to burn their fingers on oven doors anymore!

Besides the technological side benefits, basic research is necessary because it responds to the fundamental human drive to reach new forms and levels of knowledge about the world. Long ago, human beings recognized the need for maps to find sources of food in keeping with seasonal changes. This nomadic way of life, governed by bare survival instincts, may belong to the past, but humankind keeps on exploring in order to amass further wealth and knowledge. Improving and expanding maps represents one example—among many others—of how research benefits society as a whole. Scientific insight represents an integral part of culture. The efforts of researchers enrich human knowledge in general, which helps to stem tides of ignorance and violence. Education leads to happiness.

A better name for researchers might be "finders." Often I tell the students at the schools I visit that we're explorers. In fact, this view matches up with the concrete realities of my profession. By definition, explorers seek out and examine unknown terrain, come what may; the unknown draws them forward. As they proceed, they change formless matter, or intangible energy, into structured patterns of knowledge, which they then share with everyone else. And as soon as one mission's complete, they set off for new adventures.

Light as a Wave

Like sound or the sea, light displays the properties of a wave: it can be reflected, refracted, diffracted, and so on. Astronomers take advantage of all these qualities to collect light in their telescopes.

Depending on its wavelength—the distance between two successive crests in its oscillation—light can assume any color in the rainbow. For instance, light that looks bluish-violet has a wavelength of 4×10^{-7} meters (m); this is twice as short as red light, which repeats at intervals of 8×10^{-7} m. Beyond certain points, light isn't even visible to the human eye. Infrared, microwaves, and radio waves have wavelengths much longer than red light; conversely, ultraviolet, X-rays, and gamma rays have wavelengths much shorter than violet light. All these waves, visible and invisible to the human senses, constitute the electromagnetic spectrum. To determine the light spectrum of a celestial object, astronomers attach a spectrograph to the telescope. A spectrograph is usually

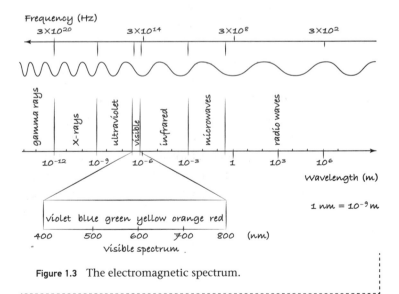

Figure 1.3 The electromagnetic spectrum.

(continued)

made up of prisms or diffraction gratings, that is, instruments for splitting the light the object emits into its various constituent colors. Although they display different hues, all electromagnetic waves—from gamma rays to radio waves—travel through space at the same speed, which is faster than anything else in the universe: light can travel 300,000 kilometers (km) in just one second.

The earth's atmosphere doesn't absorb all electromagnetic waves in the same way. It allows visible light to pass though—which is a good thing; otherwise we wouldn't see much at all. Astronomers call this range the "optical domain." Our atmosphere also lets part of the radio spectrum through: wavelengths from about around 1 mm to 10 m. Telescopes located on the earth's surface are able to collect both optical and radio waves. The atmosphere absorbs most other electromagnetic waves (X-rays, ultraviolet, infrared, and so on); to detect them, we need telescopes in space.

The Sky in Relief: Adding a Third Dimension

Almost everyone has had a go at amateur "space cartography": on hot summer nights, people contemplate the starry heavens and try to tell others about something up there that has caught their eye. The challenge is to make another person actually find the right star—just one of hundreds of lights the naked eye can see. The magical moment risks collapsing into a confused jumble of instructions. You may have regretted a lack of training on a similar occasion; it would all be much easier for an astronomer, who can specify the parts of the sky to which stars belong. Needless to say, you're not alone. Countless others experienced the same difficulty, and they recognized the value of dividing the heavens into easily identifiable regions: constellations, that is.

"Celestial vault" is the name for the hollow sphere, of inde-
terminate dimensions, surrounding the earth; all the universe's
luminous objects seem to be attached to it. Constellations are stars
that are close to one another on this surface; grouped together by
human judgment and invention, they present evocative shapes.
Although the recognized constellations vary from one civilization to the next, the International Astronomical Union has
officially divided the celestial sphere among 88 constellations,
so that each part of the sphere belongs to just one of them. To
achieve even greater precision, professional astronomers picture
the celestial sphere as a kind of grid, with large, imaginary circles
passing through the two celestial poles in one direction, and,
in the other direction, circles parallel to the earth's equator; this
design corresponds exactly to the arrangement of meridians
and parallels crisscrossing our planet. By means of this grid, any
point may be specified along two axes. Thus, in the system of

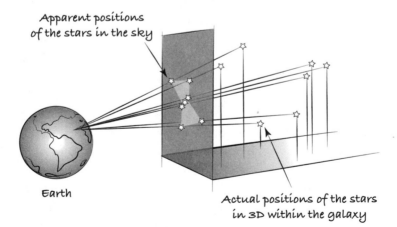

Figure 1.4 Views of Orion, in the night sky and in 3D.

so-called equatorial coordinates, a given star's direction may be described in terms of two angles: its declination, which is the analog of latitude on Earth, and its right ascension, which corresponds to terrestrial longitude.

But whether we use precise coordinates or more approximate constellations, our efforts face a major challenge: are celestial objects that appear very close to each other on the celestial vault—say, two stars belonging to the same constellation—in fact physically proximate in the universe? Not necessarily! On this score, we confront an all but insuperable obstacle: switching from a two-dimensional system of location (on the surface of the celestial sphere) to a system in three dimensions. The third dimension is depth—the distance separating us from the object we observe in the sky. In our everyday surroundings, we have no problem assessing the distance of familiar objects. Our brains automatically generate a sense of depth. The task is easy enough. In the first place, we're aware of a given object's actual dimensions; comparing our perceptions with the object's true size lets us know how far away it is. Second, inasmuch as the item in question belongs to the same order of size as we do, when our brain captures two points of view—through each of our eyes—in the same environment, it can evaluate the object's relative distance against a fixed background. This is the principle of triangulation.

But determining the distance of luminous objects "stuck" to the sky is a different matter altogether. Indeed, often we don't know how large a celestial body really is. Planets, stars, and galaxies come in all sizes, and it's hard to assign standard dimensions. What's more, these objects lie so far away from us that they frequently just look like dots. Finally, even when we use "eyes"—telescopes, that is—located at opposite ends of the earth, our planet is still much too small, in terms of the astronomical distances separating us from what we observe, for triangulation to help us out. When mapping the universe, a fair amount of ingenuity is thus required to surmount this difficulty and gauge the third dimension or depth. Methods vary in keeping with the scale of the map to be made, and the same holds for the accuracy of distances calculated. As a rule, the greater the distance is, the less reliable measurement will prove. It should be clear, then, that we cosmographers stand before a challenge of prodigious dimensions!

In fact, it is much easier to determine this third dimension with heavenly bodies that are close to us. Since the days of ancient Greek civilization, both the size of the sun and moon and their distances from Earth have been the object of calculations. Already in the third century BCE, Aristarchus of Samos was using relatively elementary geometry to figure out the distance between our planet and its moon; in his estimation, it was forty times the earth's radius (in fact, the correct number is sixty). He did the same for the distance between the earth and the sun, estimating that it's twenty times greater than the space separating the earth and the moon (it's actually four hundred times greater). Even though the ways angles were measured back then left much to be desired, his basic methods were correct.

In our own time, the distance between the earth and the moon is known with greater precision than any other astronomical measurement: astronauts in the Apollo program installed mirrors on the lunar surface. If you find yourself traveling in the vicinity of Grasse, up on the Calern Plateau in France, you might see a green laser in the sky, coming from one of the domes at the Côte d'Azur Observatory. Why are astronomers shooting at the moon? Are they trying to fight off alien invaders? Don't worry, there's no threat. The moon is home only to vast deserts of rocks, sand, and fine, abrasive dust—plus the small mirrors the astronauts have installed there, which are less than a meter across. At regular intervals, the laser sends up little bursts of light in a very narrow beam (at least when it starts out). The mirrors reflect the light, and it comes back from the moon to the observatory. It takes about two and a half seconds for the journey back and forth. By measuring exactly how much time has passed between transmission and reception, scientists are able to calculate the distance between our planet and its only natural satellite, down to the nearest millimeter!

Early Distance Measurements

The principle of triangulation—using telescopes located at significant distance from each other on the earth's surface—was first successfully applied to astronomical measurement in 1672. Jean-Dominique Cassini (1625–1712) and Jean Richer (1630–1696) determined the distance between Earth and Mars by simultaneously observing the latter's position against fixed background stars from two different points. Cassini worked from the observatory in Paris, and Richer conducted his observations in Cayenne (French Guiana).

But methods that work in the case of Mars, a planet that is close to Earth, are impracticable when the task is to measure the distance to faraway bodies like stars. Even if telescopes were positioned at diametrically opposite points on the earth, the space separating them still wouldn't be big enough, compared to the distance from a star, for the difference between the two sites of observation to make a difference. In 1838, the German astronomer Friedrich Bessel came up with a means of setting enough room between two telescopes: he waited for the earth to move!

Bessel chose a star in the constellation Cygnus, 11 light-years from our planet (as he determined in the course of the experiment), and then waited for six months, until the earth had made half its annual trip around the sun. Then, he located his star again, which had moved even farther away with respect to the fixed background of stars. Comparing the two views in terms of the angles of observation did the trick. And all it took was one telescope! Bessel had invented the parallax method. This approach takes a fair amount of patience and attention to detail: the distance between the two observation points—300 million kilometers, or twice the distance between Earth and the sun (two AUs, or astronomical units)—is still a million times smaller than the distance between Earth and another star; the accuracy needed for proper angle calculation is less than one-thousandth of a degree.

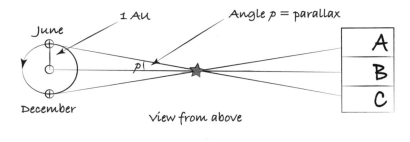

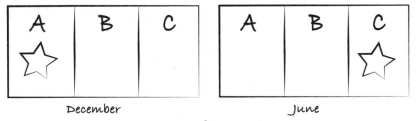

Figure 1.5 Illustration of the parallax method, after Bessel.

Astronomers still use the parallax method, but only for nearby stars. In this way, our distance from several thousand of them has been determined simply by means of telescopes here on Earth. The Hipparcos satellite launched by the European Space Agency (ESA) in 1989, which produced measurements without atmospheric disturbances, led to a veritable explosion of recorded distances during the 1990s: more than 100,000 stars were measured with a high degree of accuracy. Since then, Gaia—a new charting program that has had even greater success—has been under way. All the same, we only know a tiny fraction of the hundreds of billions of stars constituting our galaxy. Alas, there's no hope of measuring extragalactic space by means of the parallax method. But if that's the case, how did astronomers manage to discover that there are luminous objects so far away they don't even belong to our galaxy?

Units of Distance in Astronomy

Ever since the French Revolution, the international unit for length has been the meter. This measure isn't suited for the distances relevant in astronomy. When stellar systems are at issue, we use the astronomical unit (AU)—the distance between the earth and the sun, that is, some 150 million km. Otherwise, the unit astronomers usually employ is the parsec (pc), which refers, in abbreviated form, to parallax and arcsecond. One parsec represents the distance to an object calculated, by means of the Bessel method, to lie at a parallax of a single arcsecond, that is, at a remove of 0.0000278°. One parsec corresponds to about 30 trillion m. The diameter of our galaxy, the Milky Way, measures approximately 30 kiloparsecs: almost one quintillion km.

In this book, I've opted to put distance in light-years (ly); this unit, corresponding to about one-third of a parsec, is in broader

(continued)

(continued)

circulation, of course, and therefore easier to appreciate. Light travels about 300,000 km per second (km/s), which is nearly the distance between the earth and the moon (on average, that distance comes out to 384,000 km). In other words, light reflected by the moon takes just over a second to reach the earth, and vice versa. We say the moon is located one light-second away from Earth. Sixty seconds make one minute, sixty minutes make an hour, twenty-four hours make a day, and there are about 365 days in a year. The moon, then, is located some forty billionths of a light-year away from the earth. Viewed in these terms, the distance is tiny! The light-year is a huge unit of measurement.

By the same token, if we divide distance by speed, we can observe that the light emitted by the sun, which is located some 150 million km away, takes 8 minutes to reach our planet. The sun is eight light-minutes away from Earth, about fifteen millionths of a light-year. When we move to the next order of magnitude—that is, when talking about space beyond solar systems—the light-year

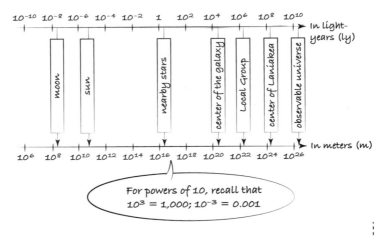

Figure 1.6   Representative distances in the universe.

(continued)

becomes a very useful unit. For instance, Proxima Centauri, the star nearest to our own solar system, is located 4.3 light-years from Earth. We can see about 6,000 stars with the naked eye, and many of them are relatively close, a few dozen light-years away. When we look at stars in the sky, our eyes are taking in light that has been traveling through space for decades. The Milky Way measures about 100,000 light-years in diameter; this means that it will take 100,000 years for a particle of light to cross from one end to the other. Light that comes from galaxies other than our own is even older (figure 1.6 provides a few representative extragalactic distances in light-years).

The farther away we look, the more we are staring into the past. The first particles of light emitted by the primordial soup that first composed the universe date from 13.8 billion years ago. Since then, the objects that produced them have moved far away from us and now are located at a distance of some 50 billion light-years. Our telescopes aren't able to capture light coming from objects over 50 billion light-years away, because it hasn't reached us yet! Our view, and the size of the observable universe, extends just 46 billion light-years.

The Great Debate

The idea that some celestial objects are located outside our Milky Way was proposed in the mid-eighteenth century. The British scientist Thomas Right and the German philosopher Immanuel Kant reasoned as follows: telescopes enable us to observe "nebulae," whose elliptical shape suggests that they comprise groups of stars similar to the one to which our own sun belongs. In the process, Kant introduced the concept of "island universes." As when Copernicus and Galileo had proposed a heliocentric system instead of a geocentric one, many contemporaries hesitated to follow this conceptual leap: the idea that our galaxy might be similar to countless others.

Supporters of Kant's theory included Germano-British astronomers Caroline and William Herschel. A brother-and-sister team, the Herschels worked together to map the Milky Way for the first time. The wager proved difficult because they located the sun at the center of the galaxy. It's hard to hold it against them, though: mapping our galaxy as a whole was like trying to paint a self-portrait without ever having seen one's reflection in a mirror, or even another human being! What's more, even now we have any number of magnificent photographs of other galaxies—some of which are quite remote—but we only have artist's impressions of our own Milky Way. In order to obtain an actual picture, the photographer would need to occupy a position hundreds of thousands of light-years away! At any rate, even if William Herschel didn't draw the most faithful representation of our galaxy, he was still an outstanding observer. His achievements include discovering the planet Uranus in 1781 and cataloging more than 2,400 nebulae. When, at long last, he managed to pierce the veil of one of them, he expected to see thousands of stars—and found only one (it was a planetary nebula with a dying star). He concluded that nebulae were not "island universes" after all; for some time afterward, the matter received no further attention.

The controversy came to a head in April 1920, when two American scientists met to debate the issue, with hundreds of their colleagues and peers in attendance. Harlow Shapley, director of the Mount Wilson Observatory, advanced the view that the universe is limited to our own galaxy, which is vast but stands alone. Heber Curtis, who worked at the Lick Observatory (also located in California), argued in support of the theory of "island

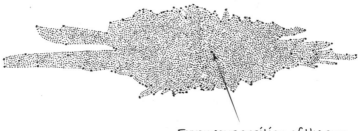

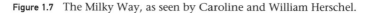

Erroneous position of the sun,
too close to the galactic center

Figure 1.7   The Milky Way, as seen by Caroline and William Herschel.

universes." Subsequent research has shown that Shapley was mistaken; all the same, his investigations produced major breakthroughs, both in terms of method and results. Indeed, Shapley is the scientist who calibrated the period-luminosity relation of Cepheids (stars that pulsate radially). In doing so, he determined the distance of star clusters from our galaxy, whose nature could then be defined. As such, he deserves credit for putting the sun at the edge of the Milky Way.

Absolute Luminosity, the Key to Measuring Distance

That said, what does it really mean to "calibrate the period-luminosity relation of Cepheids"? Cepheids are giant stars at the end of their lives. Their defining trait is that they produce bursts of brightness exhibiting slight but regular variation over time, because their nuclear "fuel" is running low. The time of a cycle (also called a period) varies from one star to the next; it can last for a few days or several months. The name *Cepheids* was coined because a star of this type is found in the Cepheus constellation. In fact, they might have been given a completely different name.

At first, they went by the title of "variable stars." Polaris, the North Star, is a prime example. The period-luminosity relation captures the rate at which a star's luminosity increases in the course of its cycle: when one Cepheid is brighter than another one, it "twinkles" more slowly.

Henrietta Leavitt first observed this phenomenon, which represents a major turning point in our efforts to gauge the depth of the sky. Beginning in 1907, she scrupulously examined countless photographic plates of the Small Magellanic Cloud. The Small and Large Magellanic Clouds are satellite galaxies of the Milky Way. Visible to the naked eye only in the southern hemisphere, they were remarked by the explorer Ferdinand Magellan as he circumnavigated the globe during the early sixteenth century. One hundred years ago—when Leavitt was conducting her research—no one knew that the Magellanic Clouds are located outside our galaxy. She noted the relation between the cycles and the apparent brightness of the Cepheids forming the Small Magellanic Cloud. In turn, Shapley contributed to our understanding by measuring the periodicity of a Cepheid located at a distance already calculated by other means. In this way, he was able to determine its absolute luminosity. Using Leavitt's work, he deduced the relation between the period and absolute luminosity: this is what it means when we speak of "calibration."

This method is still used today, but only for determining our distance from nearby galaxies. Indeed, there aren't many variable stars, and those that do exist aren't very powerful. Researchers start out by locating a Cepheid in a galaxy and measuring its pulsation period; then they use this information to deduce the star's absolute luminosity (its light output in watts) by means of the period-luminosity relation. Comparing absolute luminosity

Henrietta Swan Leavitt

Figure 1.8

American astronomer Henrietta Swan Leavitt (1868–1921) discovered the relation between the absolute luminosity and the period of pulsation for Cepheid stars by carefully studying photographic plates. After earning advanced degrees at Radcliffe, she worked under Edward Pickering (1846–1921) at Harvard University. Other famous astronomers belonged to this research group—including Annie Jump Cannon (1863–1941), who devised a system of classifying stars that is still used today. Henrietta Leavitt was deaf and suffered from poor health. All the same, her mettle and sound intuition enabled her to achieve excellent results. She was eventually promoted and made head of the photographic stellar photometry department. Before long, researchers used her innovations to measure the distance of galaxies outside our own for the first time. In doing so, they discovered that the universe is expanding. Edwin Hubble, who demonstrated this process, is said to have expressed the view on many occasions that Henrietta Leavitt deserved to receive the Nobel Prize for her work.

to the star's apparent brightness (in other words, the luminous flux registered on Earth, in watts per square meter) yields the galaxy's distance. During the 1920s, at Mount Wilson Observatory, Edwin Hubble, working with the largest telescope of that time, used this method to determine the distances of "extragalactic nebulae" for the first time in history. He confirmed the hypothesis Kant had advanced two hundred years earlier: many other "island universes" exist.

The great adventure of modern cosmology could finally begin. Before long, Hubble made other important findings. In particular, he discovered a simple equation that enables us to estimate the distance of remote galaxies on the basis of their movement; his approach is still widely employed. But the details will have to wait until later. Now it's time for me to take the stage!

Absolute Luminosity and Apparent Brightness

The absolute luminosity of a celestial body corresponds to the luminous power it emits. This value is expressed in watts (W), just like the power of the light bulbs we use in our homes. But whereas the order of magnitude for a light bulb's luminosity is in the range of 100 W (or 1×10^2 W), the order of magnitude for the luminosity of a star like the sun is 1×10^{26} W; for a galaxy, it's 1×10^{37} W. As anyone can readily observe, moving away from a source of light makes the light perceived much less intense. This is the case because the power emitted by the bulb—with an absolute luminosity of 100 W—is dispersed across a broad expanse: to wit, the area of a sphere whose radius is equal to the distance between our eyes and the bulb. For instance, if one stands at a meter's distance from the light source, the "apparent brightness" reaching one's eyes is a flux at 8 W/m^2; and if one stands 3 m away, the flux is less than 1 W/m^2.

(continued)

Figure 1.9 Absolute luminosity and apparent brightness.

Accordingly, a simple relation holds between these two quantities, apparent brightness and absolute luminosity; it's a matter of squaring the object's distance from the earth. Astronomers use this simple equation as follows: with a telescope, they measure the apparent brightness of a star (a practice known as photometry); then, once its absolute luminosity has been determined, it's easy to calculate its distance from the earth. Unfortunately, however, light bulbs come with indications of how bright they are; in contrast, the absolute luminosity of celestial bodies (galaxies, for instance) isn't written on the packaging. Ultimately, our efforts to sound the depths of the sky still depend on guesswork about the absolute luminosity of celestial bodies.

My First Steps into the Unknown

In 1992, at the age of twenty-two, I arrived at the Lyon Observatory to complete my master's degree by working on the team supervised by Georges Paturel. During the 1970s, Paturel had begun collecting data that would enable him to calculate our distance from galaxies in the Local Universe. This work occurred in collaboration with colleagues at the Paris Observatory (in Meudon), Lucienne Gouguenheim and Lucette Bottinelli; for the most part, they conducted their own radio-astronomical observations from Nançay in Sologne. In 1983, Paturel embarked on a massive project. He decided to bring together, in digital format, all the information he and others had obtained. In other words, he created the world's first extragalactic database, LEDA (Lyon-Meudon Extragalactic Database), subsequently renamed HyperLEDA. Along with NED (NASA Extragalactic Database), it represents a resource of worldwide significance.

Discovering these new, extragalactic worlds, I immediately felt their charm. As fast as I could, I devoured every article I could find about large structures in the Local Universe. As my readings proceeded, I felt myself being carried far beyond the Milky Way. I pictured the tiny corner of the universe that cosmologists have dubbed the "Local Group," where our galaxy and its two closest neighbors—Andromeda and Triangulum—display their finery like graceful queens at court, with a host of dwarf galaxies in their train. I also recognized that Andromeda's realm stands dangerously close to our own!

The dwarf satellite galaxies made my head spin; they seemed to "orbit" the Milky Way, much as planets revolve around stars. And as soon as my attention turned from the Local Group, I found myself teetering at the edge of the immense Local Void. In my mind's eye, I was practically flying at the speed of light. But even if one travels this fast, the voyage risks taking quite some time... I've already gone 50 million light-years when I come to a halt, awestruck before a magnificent spectacle: more than a thousand huge galaxies crowding together make this part of the sky look like a busy highway at rush hour. I find myself in the Virgo Cluster, which is much more imposing than the small court of our Local Group, with its three queens. Stepping back even more, I notice a few isolated galaxies—country cousins, as it were. I feel as if I'm looking at the headlights of cars racing down the streets connecting different parts of a great metropolis. Fornax, Ursa Major, and others combine to form the celebrated "Local Supercluster." It seems to be trying to extend its filaments as far as the Hydra-Centaurus Supercluster; they brush so close they practically touch. At an even greater distance, the Perseus-Pisces, Coma, and Pavo-Indus Superclusters encircle our Local Universe with a vast web hundreds of millions of light-years across. I'm getting a taste for the dizzying heights.

I can only assume that the first astronomers to map this part of the universe explored space in the same way—travelers like Gérard de Vaucouleurs, the celebrated astronomer from France who emigrated to the United States. In 1953, he published proof of a "local supergalaxy," which was later renamed the Local Supercluster. (It kept this title until our own discovery of Laniakea.) When studying the distribution of galaxies, we still use his supergalactic coordinate system; its axes connect at this same Local Supercluster (which is relatively flat).

In the 1960s, Donald Shane and Carl Wirtanen conducted the first grand-scale galaxy counts in two dimensions. Distant galaxies, it seems, bunch together in vast structures that, unsurprisingly, are given the names of the constellations behind which they appear: for instance, "Hydra Supercluster," or "Centaurus Supercluster"... These initial observations did not identify galaxies

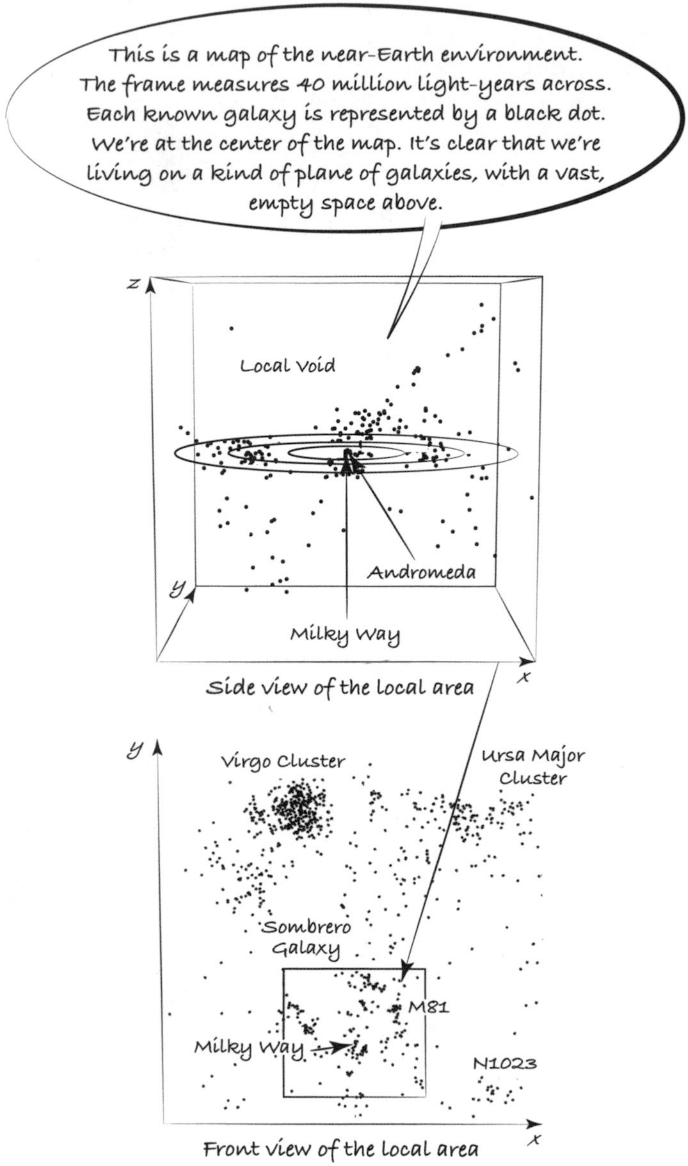

This is a map of the near-Earth environment.
The frame measures 40 million light-years across.
Each known galaxy is represented by a black dot.
We're at the center of the map. It's clear that we're
living on a kind of plane of galaxies, with a vast,
empty space above.

z

Local Void

Andromeda

Milky Way

y

x

Side view of the local area

y

Virgo Cluster

Ursa Major
Cluster

Sombrero
Galaxy

M81

Milky Way

N1023

x

Front view of the local area

Figure 1.10   Close-ups of the Local Universe.

individually; researchers just counted them in the squares of the grid covering the celestial vault. In 1986, Valérie de Lapparent, at the Paris Observatory, collaborated with American researchers Margaret Geller and John Huchra to produce the first large-scale, three-dimensional map; shaped like a fan, it covers part of the celestial vault. In the course of their work, they discovered the first giant cosmic structure, which they named the "Great Wall." Its immense dimensions are given in millions of light-years (Mly): 500 Mly (length)×200 (height)×15 (width)! Even more remarkably, they demonstrated that the Local Universe is arranged in "bubbles": empty spaces surrounded by partitions formed by galaxies—like the Great Wall.

Figure 1.11 Valérie de Lapparent delivering a talk on large galaxy structures, including the Great Wall, at a 2009 conference.

Weaving My Cocoon

My first task was learning how to homogenize the data in LEDA, which was often expressed in incompatible terms. After all, the observations it brought together had been made by an array of research groups using different instruments and analyzing their findings by different methods. For instance, cosmologists frequently use the Hubble constant (which I'll discuss at greater length in later chapters) in order to estimate the distance of a galaxy. Yet the value of this constant is still not well measured, and researchers don't necessarily use the same figures—which means that estimates can vary greatly. Needless to say, the simplest way to avoid troubles would have been to conduct observations on our own; in that case, the parameters of measurement would be clear. But time for making observations is limited. Back then, a given research team could measure only a few dozen galaxies a year—maybe a few hundred, but not more. It was just a drop in the ocean of the universe...

When research groups coordinate their efforts, it's much easier to obtain details about the methods employed; this helps, when comparing others' findings to one's own, to standardize measurements across the board and improve their reliability. Unfortunately, there's often no communication at all between research teams. When this is the case, it's necessary to wait until observational data has been analyzed in full and the results have been reviewed and published in professional journals. Sometimes, there's no choice but to hang on until information falls into the public domain. It's comparable to the situation when, say, a lab develops a new chemical formula for a medicine and reserves use for itself. In the field of astronomy, no standard legal

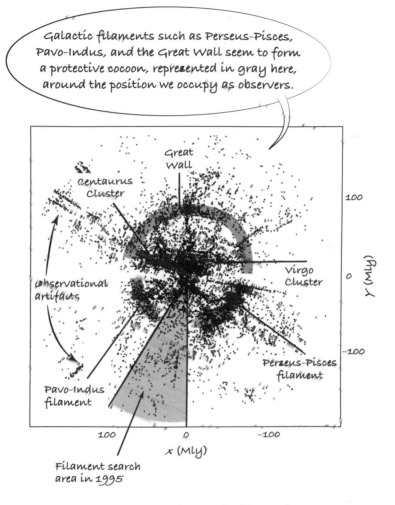

Figure 1.12 One of my first maps showing the "Cocoon."

duration holds for data; how long it takes to be shared varies from one telescope and observatory to the next.

Formatting all available measurements isn't the whole story, either. One also has to know how to identify data that's corrupted or imprecise, then eliminate it.

Once I'd completed that part of the job, I could finally go about making my first maps. To identify these structures more fully, I wrote an algorithm to animate the maps and show them from different angles. It was my first foray into visualizing digital information. At the beginning of the 1990s, exchanging images around the world wasn't as easy as it is today. I remember that when we wanted to share the first of our extragalactic maps with other specialists in the field, we took photographs of our computer screens, which we then had printed on plastic transparencies, so they could be used on overhead projectors at conferences. Even though they were of mediocre quality, these transparencies made fruitful exchanges possible and, as such, played a crucial role in the identification of the principal galaxies. In particular, Gérard de Vaucouleurs's positive assessment of my early maps gave me a sense of confidence and motivation that inspires me to this day.

The distribution of galaxies I sketched back then brought an unexpected structure to light. The Virgo Cluster, in whose proximity our Local Group is situated, occupies the middle of a kind of gigantic bubble: the Great Wall, the Perseus-Pisces Supercluster, and the Pavo-Indus Supercluster seem to form a ring around us. If one discounts a few empty spaces that set them apart from each other, they look like a smooth, three-dimensional structure surrounding us and cutting us off from the rest of the universe. Thus, in honor of Lyon's venerable tradition in the silk trade, Georges Paturel and I named my first discovery the "Cocoon."

2 In Pursuit of the Great Attractor

Or how, following the discovery of the way our galaxy moves in the universe, a team of American cosmologists found the way to the Great Attractor by creating the first dynamic maps.

The Universe in Motion

When we admire the skeleton of a tyrannosaurus in a natural history museum, the bones don't hold our attention. Before long, we're picturing the whole animal, muscles tensed and ready to strike; then, almost as quickly, it's off—and heading our way! Scenes like this illustrate how visualizing movement gives meaning to an otherwise empty structure. In the same way, measuring tectonic shifts led scientists to recognize that India never belonged to the Eurasian Plate, and that the Himalayas emerged when two continental plates collided. Identifying patterns of movement sheds light on their effects, as well as their causes. Thus, "visible" tectonic movement reveals attendant energy that's invisible, hidden in the depths of the earth. The same rule holds for galactic structures. Access to the third dimension—the depth of the sky—gives us a glimpse of galaxies distributed in space, enabling us to see areas where they group together and regions they leave behind. But to understand the reasons why such organization occurs, and how it will evolve, we need to measure the motion of galaxies.

Charting Everything That Moves

Recording motion represents an incredibly useful tool for understanding the world. It might seem to serve no purpose in cartography, which, after all, generally focuses on fixed positions. But all the same, movement plays a vital role in a wide range of cases. Although many examples—and highly varied ones—could be given, I will provide just two. They don't relate directly to the dynamic maps of the universe that I make; instead, they concern parts of the earth that are dear to my heart, where I've spent some wonderful years as an astrophysicist.

Approximately 50,000 years ago, the original inhabitants of Australia followed a small stream trickling down a rocky monolith up to hidden wellsprings of precious water—between meandering crags that cast a sacred glow. No map was drawn to preserve the memory of these sources of life; instead, the path was sung or danced. Aboriginal peoples also oriented themselves by looking at the nighttime sky. In contrast to modern, Western astronomy, which names clusters of stars, they had "constellations" referring to dark regions of the Milky Way—clouds of dust and gas.

Some eight hundred years ago—far, far to the east of Australia—Polynesians traveled the Pacific Ocean in search of new islands. They did so by drifting along with currents, on the lookout for signs of plant or animal life: floating debris such as coconuts, or birds that had wandered away from shore. They also used the apparent motion of stars for guidance. Here, too, the names of constellations and ocean currents were transmitted orally. Later waves of colonization wiped out the tradition. Recently, however, the Polynesian art of navigation has been rediscovered and the names for heavenly bodies in Hawaiian have been revived.

Great conceptual advances in the field of astronomical cartography often start with observing and recording phenomena of motion. In ancient Greek civilization, around 150 BCE, Hipparchus of Nicaea gauged the distance between the earth and the moon by closely watching the latter's movement in the former's shadow—or, more precisely, by comparing the duration of a lunar eclipse with the time the moon took to circle the earth—in light of work conducted by his predecessors (Aristarchus of Samos and Eratosthenes of Cyrene). Four hundred years ago, Galileo (1564–1642) observed the motion of four of Jupiter's satellites with telescopes he made himself, which could enlarge images thirty times; these simple tubes containing a series of glass lenses would be deemed rather crude by today's standards. But by recording the satellites' movement in a series of drawings made at different times, Galileo recognized that they were revolving around Jupiter. If small objects tend to orbit larger ones, he concluded, the earth must orbit the sun—not the other way around! Recording periodic movement added fuel to the fire Copernicus had started with his heliocentric model—and caused considerable problems for Galileo. After all, this new view of the world cast doubt on Catholic dogma.

At the same time, German astronomer and mathematician Johannes Kepler (1571–1630) demonstrated that planets orbit the sun along elliptical, almost circular, trajectories. He did so on the basis of painstaking studies of their movements conducted by the incomparably astute Danish observer Tycho Brahe. These first astronomical observations, which shed new light on the

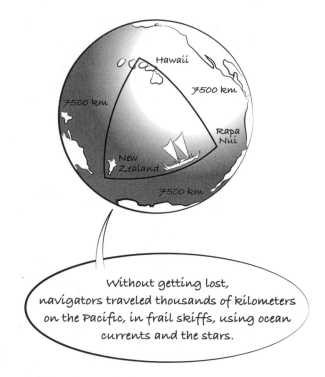

Figure 2.1 Polynesian navigation.

most fundamental physical phenomena, marked the beginning of astrophysics.

Expanding Space

One hundred years ago, when the field of observational cosmology was just starting out, more attention was paid to locating nebulae—whether inside or outside our galaxy—than to following their movements. All the same, American astronomer Vesto

Slipher (1875–1969) set about measuring the wavelength shift of radiation emitted by nebulae. Interpreting this shift as a velocity, the shift resulting from the Doppler effect (see "The Doppler Effect," below), he showed that certain nebulae are moving at a speed of hundreds of kilometers per second! His discovery also added credence to the idea that nebulae are extragalactic bodies. Behind the seeming immobility of the stars, a bustling world lies hidden…

In 1929, Edwin Hubble continued in this vein by measuring galaxies' velocity (in terms of the Doppler shift) and distance (in terms of Cepheid variation and galaxy luminosities). Reading Hubble's paper "A relation between distance and radial velocity among extra-galactic nebulae" is a healthy reminder of how much clearer things can become with decades of hindsight. For example, although the Cepheids lie at the foundation of Hubble's distance scale, the distances to most of the objects in his article were not determined by Cepheids themselves, but by the brightest stars in galaxies or by the luminosity of the galaxies themselves. He was surprised to see that faraway galaxies seem to be moving away from us systematically: they're "receding." More remarkable still, the more distant a galaxy is, the faster it's moving away! Hubble had discovered the law that now bears his name:

$$V_{\text{expansion}} = H \times D.$$

Hubble recognized that the velocity of a galaxy moving away from us is directly proportional to the distance D separating us from it. This proportionality constant is denoted by the symbol H, for Hubble. Thereby, Hubble demonstrated beyond all doubt that the universe is in the process of expanding. The observation has played a key role in the success of the Big Bang model—a theory that's hardly intuitive, by the way: even

Light as a Particle

Light possesses the properties of a wave (as we saw in the previous chapter), but it may also be viewed as a particle without mass: a photon. In 1905, Swiss physicist Albert Einstein showed that the energy carried by a photon is directly proportional to its frequency and, therefore, inversely proportional to its wavelength. Accordingly, a violet photon carries twice as much energy as a red

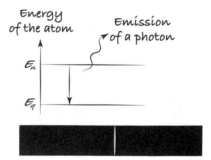

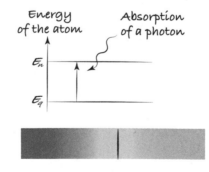

Figure 2.2 Spectral transitions in the course of a photon's emission and absorption. (Also plate 2.)

(continued)

---

photon, because its wavelength is half as long. What's more, the photons of large radio waves are much less dangerous than X and gamma rays, which have shorter wavelengths and higher energy levels.

Light particles can interact with matter, particularly atoms. Indeed, quantum mechanics tells us that atoms have multiple, highly specific energy levels, corresponding to the only positions electrons can occupy in an atom. When a "hot" (excited) electron passes from a higher energy level to a lower one—that is, when it "deenergizes"—it emits a photon; this photon's wavelength—its color, in other words—corresponds precisely to the energy lost during the transition between the two levels. Figure 2.2 shows an atom passing from energy level E_n to E_q, which is lower, thereby causing the emission of a yellow-colored light particle. Inasmuch as energy levels depend on the atom in question (a hydrogen atom has different energy levels than a calcium atom, for instance), every atom emits specific colors when it deenergizes—what we call "spectral lines." A given chemical element always produces the same "emission spectrum"; thus, the spectrum of hydrogen always differs from that of calcium. Conversely, a "cold" (unexcited) atom illuminated by an external light source is able to absorb the light of a specific color, giving it exactly enough energy to attain a higher level. The missing light (missing because the atom has absorbed it) appears as a black line on the spectrum. In this way, each chemical element presents uniform absorption and emission lines at the same wavelengths.

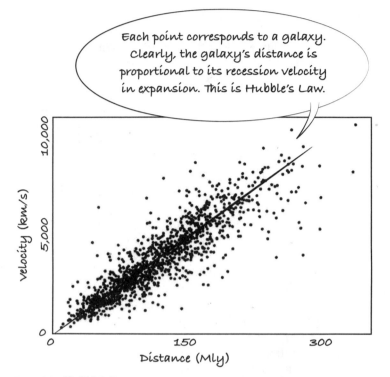

Figure 2.3 Hubble's Law.

Einstein failed to consider its possibility. One way of picturing the phenomenon is to bake a loaf of bread (the universe) with raisins (galaxies). As the bread rises, the raisins move apart—and at a rate that increases in proportion to their initial distance from each other. If you don't bake, there's another trick you can try at home for visualizing the process of expansion. Put small stickers on a balloon and inflate it; you'll see how the "galaxies" move away from each other as the "universe" expands.

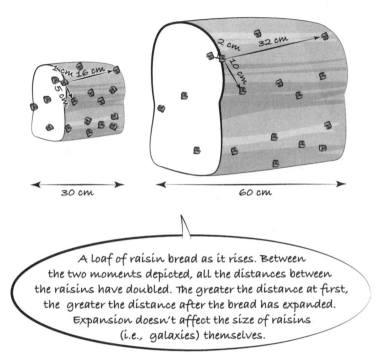

Figure 2.4 The expansion of raisin bread.

Hubble not only demonstrated that the universe is expanding; he also gave cosmographers a new tool for determining the distance of galaxies. Indeed, measurements of velocity in terms of the Doppler effect can be made quickly and with great precision; with this information in hand, researchers have no trouble calculating distances—provided that they bear in mind the proportionality constant H and disregard motion that isn't due to expansion. Cosmographers speak of the *redshift* method.

The Doppler Effect

When a fire engine comes toward us, we hear a sound more high-pitched than the "actual" siren we would hear if the vehicle weren't moving. Likewise, the sound becomes more low-pitched as it moves away from us. All waves display this property, which is known as the Doppler effect in honor of the man who discovered it, Austrian physicist Christian Doppler (1803–1853).

As the source approaches the observer, the waves are compressed: they arrive at a higher frequency and with a shorter wavelength. Conversely, when the source moves away from the observer, the waves stretch out: now, they arrive at a lower frequency and with a longer wavelength. The effect grows more pronounced when the speed of the source is significant in comparison with the speed of the wave. Accordingly, in our day-to-day lives, we register this phenomenon more readily with sound than with light, which travels too quickly. In the field of astronomy, on the other hand, the Doppler effect occurs only with light. This is the case because, first, sound cannot travel through the interstellar and extragalactic void. Second, the movement velocity of the celestial bodies emitting these waves is so great that it is only rarely insignificant, even compared to the speed of light. When a luminous object moves toward us, the light it emits is compressed and its wavelength decreases; put in terms of the visible part of the electromagnetic spectrum, it moves toward the color blue, hence the name "blueshift." By the same token, when a celestial body moves away from us, the waves spread out and move toward the red end of the spectrum; this is called "redshift" (see figure 2.5).

When studying distant galaxies, we always observe some kind of redshift as a result of the universe's expansion. Astronomers use a spectroscope to measure the shift of emission lines and absorption lines relative to their values in a state of "rest." Then, using the mathematical formula that accounts for the Doppler effect, they calculate the velocity of the source. The biggest problem with this method is that it tells us nothing about the galaxy's radial velocity, that is, velocity in terms of the viewing direction

(continued)

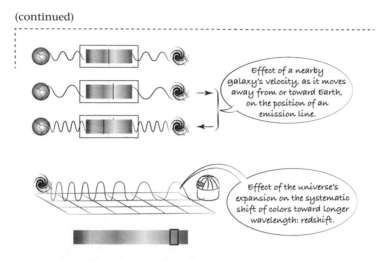

Figure 2.5 Using the Doppler effect in observational cosmology. (Also plate 3.)

between the observer and the galaxy. There's no way of knowing whether the galaxy is moving in a way other than this radial direction, since the Doppler effect doesn't occur on the axis perpendicular to the line of sight. In other words, the Doppler effect cannot tell us about the galaxy's tangential velocity.

How to Find the "Peculiar" Velocity of a Galaxy

Hubble's Law gives researchers a veritable Swiss army knife, as it has even more applications. By this means, they can show the residual motion of galaxies, once the effects of expansion are removed. A galaxy's "total" radial velocity is the sum of its velocity in the course of expansion and what's known as its "peculiar" velocity.

$$V_{\text{total}} = V_{\text{expansion}} + V_{\text{peculiar}}$$

"Peculiar" motion occurs because of the surrounding gravitational field. My research focuses on "peculiar" movements because they allow us to identify the cohesion of large structures with a greater degree of certainty than fixed maps provide. The process for determining these peculiar velocities is always the same:

1) First, by means of the Doppler effect, astronomers measure the "total" radial velocity of the galaxy.

2) Then they calculate the galaxy's distance, D, with a method other than redshift. In the previous chapter, we saw how this is done with Cepheids, but there are other methods, too. Below, we'll discuss the Faber-Jackson method for elliptical galaxies; in the next chapter, we'll look at the Tully-Fisher method for spiral galaxies.

3) In keeping with Hubble's Law, multiplying the distance D by the constant H yields the radial "recessional" velocity $V_{expansion}$ at which the galaxy is moving away from us due to cosmic expansion.

4) Finally, they subtract the recessional velocity due to cosmic expansion from the total velocity, which provides the galaxy's "peculiar" radial velocity. This residual motion, caused by the gravitational environment of the galaxy, is independent of motion that occurs through the expansion of the universe.

$V_{peculiar} = V_{total} - (H \times D)$

We follow this simple procedure, which takes just a few steps, for most of the observations we make when mapping moving galaxies. Although it all seems relatively straightforward, in practice it presents any number of difficulties. For one, it's no small matter to determine a galaxy's distance by means other

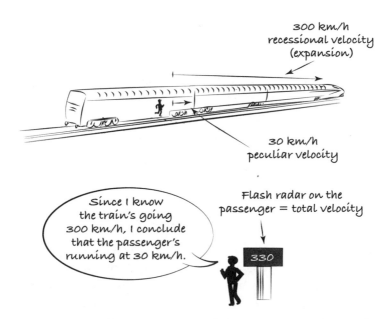

Figure 2.6 Total velocity, recessional velocity, and peculiar velocity of a passenger running inside a train.

than redshift. The most common approaches, which use the so-called Faber-Jackson and Tully-Fisher relations, yield imprecise values. What's more, the value assigned to the H constant varies from one research team to the next—a real problem inasmuch as determining the role played by cosmic expansion as precisely as possible is key to figuring out how fast a galaxy moves on its own. Indeed, peculiar velocity is often ten times smaller than expansion velocity, and this ratio only increases the farther away the galaxy is. To picture the difficulties involved, imagine trying to measure how fast someone is running down the aisle of a high-speed train, while standing on an open field and watching the train race past. It's impossible to do without knowing the train's speed beforehand.

Discovering Our Galaxy's Motion through the Universe

Now that's the kind of challenge I was eager to take up—though the mission would prove even harder out in space! Come what may, I had to succeed. The stakes were high: ever since movement had entered the picture, more and more research teams had gotten to work, and they were asking more and more questions. The most exciting development, to be sure, occurred in 1964, when two engineers at Bell Labs calibrating an antenna discovered "cosmic microwave background radiation": very-low-energy electromagnetic radiation from a time long ago, when the universe was very hot, which had cooled in the course of its expansion. For making this discovery, which strengthened the theory of the Big Bang, they were awarded the Nobel Prize in Physics in 1978.

The observation involved another finding of even greater import for cartographers of the Local Universe: the frequency value of such radiation varies in keeping with where measurement instruments are pointed. This preferential direction is called the cosmic microwave background dipole. Although it's surprising to see it when one is expecting homogeneous and isotropic radiation, the phenomenon is easy to explain in terms of the Doppler effect: it's a matter of the earth's movement in relation to the universe. Once we take away the velocity of the earth around the sun, the sun's velocity within our galaxy, and our galaxy's motion toward Andromeda, we arrive at a startling result: our galaxy is traveling at the rate of about 630 kilometers per second. A speed like this is altogether normal for galaxies in large galaxy clusters, but not normal at all in an isolated setting like the one occupied by the Milky Way.

Might this be motion occurring on a much greater scale—the result of some enormous, faraway mass pulling at us? Unfortunately, the direction in which it seems we are heading (toward the Hydra-Centaurus Supercluster) appears behind our galaxy's own plane in projection on the sky. Efforts to conduct observations in this direction—beyond the Milky Way—to look for a mass responsible for our motion are obstructed by countless stars and the dust from our own galactic disc. We can't see through! To appreciate the difficulty astronomers confront, hold your forearm horizontally, about 20 centimeters from your face. Your arm is blocking a significant portion of your field of vision—just as the Milky Way does for astronomers. Needless to say, locating the mysterious attractor hidden nearby—in what's known as the "Zone of Avoidance"—is fraught...

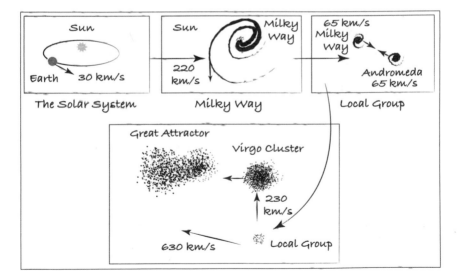

Figure 2.7   Our large-scale motion in the universe.

The Seven Samurai

In 1976, American astronomers Sandra Faber and Robert Jackson discovered a law that would enable researchers to determine the distance of elliptical galaxies by indirect means—an essential step toward calculating peculiar velocity. It's important to bear in mind that stars near the gravitational center of elliptical galaxies do not display a uniform pattern of motion: they move in all possible directions. (This is very different in the case of spiral galaxies; here, all the stars revolve in the same direction around the central bulge.) Faber and Jackson identified a relation between the chaotic movements of the central stars in elliptical galaxies— what's known as velocity dispersion—and their absolute luminosity (see "The Faber-Jackson Relation," below). With this relation in hand, they could measure velocity dispersion and calculate absolute luminosity; then, by comparing absolute luminosity with apparent brightness, they determined the galaxy's distance. Finally, by the procedure explained above (page [30]), they had everything they needed to figure out its peculiar velocity.

Faber and Jackson had worked out the first method for revealing the peculiar motion of faraway galaxies, without getting tripped up by the phenomenon of expansion that separates them from us. In so doing, they opened the way for "dynamic" cartography, which soon showed that galaxies and the larger structures to which they belong (clusters and superclusters) are subject to bulk motion, as if invisible currents were carrying them along. The observation seems fairly elementary today; at the time, however, it called into question the generally accepted view according to which the universe is more or less static at these levels. When they presented their first dynamic maps of the Local Universe at a conference in 1986, Faber and her colleagues met with some disbelief;

Sandra Faber

Figure 2.8

After earning her PhD from Harvard University, Sandra Moore Faber (b. 1944) was the first woman to join the faculty of the Lick Observatory at the University of California. Codiscoverer of the dispersion-luminosity relation for elliptical galaxies and a specialist in the formation and evolution of galactic structures since the Big Bang, she has also pioneered research on the role dark matter plays in them. An expert of optical observation, Faber was instrumental in designing the Keck telescopes in Hawaii. With a primary mirror 10 m in diameter (composed of 36 hexagonal segments), these telescopes are the largest in the world. Faber also helped to correct the spherical aberration that initially plagued the Hubble Space Telescope. Her numerous awards include receiving the National Medal of Science from President Barack Obama in 2013.

The Faber-Jackson Relation

The Faber-Jackson relation is an empirical equation that links the gap ("dispersion") between the velocities of stars near the center of an elliptical galaxy to its absolute luminosity. The velocity of stars' motion is obtained by measuring changes in absorption lines—e.g., those of calcium, sodium, or magnesium—in the visible portion of the galaxy spectrum. The accuracy of measurements poses difficulties because it's hard to know how much variation derives from the stars' rotation in the galaxy or the effects of turbulence in the "envelopes" of the stars themselves... As when measuring distance by any other means, researchers must calibrate the relation at each stage: first, in terms of nearby objects tested by other methods, then, little by little, with objects at a greater remove.

The spectrograph collects light from all stars seen through the aperture.

Velocity dispersion is measured by the size σ of an absorption line.

Aperture of the spectrograph

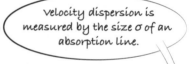

Absorption line
(sodium, calcium, or magnesium)

σ km/s

Faber-Jackson relation: $L_{galaxy} \propto \sigma^4$

Figure 2.9 Enlargement of an absorption line in an elliptical galaxy.

Our nearby extragalactic environment contains no elliptical galaxies, however. Accordingly, the calibration of the relation is not very precise. The margin of error for these distance measurements

(continued)

lies at about 30%. Often, another empirical relation is preferable: the Tully-Fisher relation, which can be used for spiral galaxies. (See the next chapter.) But even if measurements employing the Faber-Jackson relation have been infrequent in the last thirty years, the practice has just made a giant leap forward. In 2015, the Australian 6dF (Six-degree-Field Galaxy Survey) observation program published a catalog of distances for over 8,000 elliptical galaxies obtained by this method.

a specialist in the field, Amos Yahil, was so convinced they were defending a lost cause that he ironically dubbed them the "Seven Samurai" because they were committing professional suicide.

Gravitation Takes the Stage

What makes galaxies move, if not the expansion of the universe? The answer's simple: gravitation—the same force that makes the moon orbit the earth and famously caused an apple to fall on Newton's head. If we're talking about phenomena on an astronomic scale, gravitation is always the culprit. In 1984, Sandra Faber teamed up with other experts in the field (George Blumenthal, Joel Primack, and Martin Rees) to publish a major article on the topic, a key point of reference for all further research. The authors underscore the central role that "dark matter" plays in the movement of galaxies. Dark matter: it certainly sounds mysterious! The term refers to massive particles that emit no light; when luminous matter alone cannot account for galactic velocities, scientists invoke the power of dark matter.

Making a velocity map requires lengthy and painstaking observation and analysis. At the end of the 1980s, Faber assembled a team of six other colleagues to help with the project. These Seven Samurai worked together for eight years to catalog the positions, sizes, and velocities of 400 elliptical galaxies. Their massive efforts located the region that seems to be pulling our galaxy and its neighbors in a direction close to the one the cosmic microwave background dipole suggests. Remarkably, Faber's team noticed that the Centaurus Cluster, which was thought to account for the attraction experienced by our Local Group, also seems to be moving away. In other words, the dominant center is located even father off… What's more, the quantity of luminous matter visible as galaxies at this site does not match the incredible gravitational force necessary to make galaxies move this fast.

On the basis of these findings, Faber and her team proposed the existence of a massive amount of dark matter and galaxies—some more or less spherical object drawing everything toward itself. They called it the "Great Attractor." Unfortunately, the Great Attractor's location is difficult to determine with any precision, since in the sky it's close to the zone obscured by our own galaxy (remember the simulation above, holding your arm in the field of vision). In addition, it's so far away that it stands at the outer limits of what the telescopes of the day could detect. These limitations prevented the Seven Samurai from confirming their theory.

Gravitation and Other Fundamental Interactions

Gravitational attraction, or gravitation, is the force two masses exercise on each other. Everyone is familiar with it: the pull exercised by our planet's enormous mass (6×10^{24} kg) is an everyday phenomenon. But for all that, this force is by far the weakest of the four fundamental interactions governing the universe (weak and strong nuclear forces, electromagnetic force, and gravitation). Why does it play a predominant role on the astronomic scale?

(continued)

Unlike gravitation and electromagnetic force, whose ranges are infinite, the two "nuclear" forces have a spatial range limited to the size of an atom's core. As such, even if the strong nuclear force is the most powerful of all—enough to hold protons and neutrons together, thereby ensuring the cohesion of the nucleus—it ceases to exercise the slightest influence as soon as these particles are separated by even a few femtometers (that is, millionths of a billionth of a meter). At this point, the third force—electromagnetic interaction—takes over, which governs the organization of all structures from the atomic level up to the scale of human life and activity. It is at work in chemical reactions when electronic bonds are broken and formed, and in changes that occur in the physical state of matter...in a word, anywhere and everywhere an electronic charge is present. In absolute terms, it is much, much stronger than gravitation (10^{40} times stronger, in fact). But since opposite charges are drawn to each other, matter itself tends to be neutral on a large scale. This is clear enough: most of the objects we encounter every day are electrically neutral. If there's no charge, there's no longer any electromagnetic force; in consequence, the least powerful force (in intrinsic terms) is what governs large objects: gravitation.

Galileo examined gravitation on an experimental basis; in turn, toward the end of the seventeenth century, Isaac Newton devised a model for it when, according to legend, he saw an apple fall from a tree. Newton's Law provides an adequate account of gravity's effects in terms of so-called classical mechanics—that is, when the velocities of the objects at issue aren't too large. Discrepancies between theory and what may be observed experimentally emerge only when velocity approaches the speed of light (which is the domain of relativistic mechanics). In 1915, Albert Einstein proposed his theory of general relativity, which does away with such differences. Einstein didn't view gravitation as a force so much as a warp that the energy of a mass produces in the continuity

(continued)

(continued)

> between space and time. To his great dismay, the theory of general relativity implied a nonstatic universe—much like the expanding universe scientists observe today. A hundred years ago, the notion of a dynamic universe defied reason, and Einstein felt obliged to add a "cosmological constant" to his theory so the universe would remain stable. He would later speak of the introduction of this cosmological constant as the "greatest blunder" of his career.

Diving into the Southern Universe

As a budding, twenty-three-year-old astronomer, the first discovery I could pride myself in having made was the subject of my doctoral research: I wanted to find out all I could about the Great Attractor! Like the Seven Samurai, I would need to map the motion of the Local Universe in order to identify the zones responsible for points of galactic convergence. That said, the region where the Great Attractor is located cannot be seen from the northern hemisphere. I wasn't about to let that get in my way, and I set off for Australia. The lengthy voyage from Paris to Sydney, via Singapore, gave me plenty of time to think about everything I'd encounter on this new continent besides telescopes: its native peoples, kangaroos, the Great Barrier Reef... Looking back, the most striking thing—more than the assorted clichés that inspire young Europeans to seek adventure—was the contrast between life in the big city and life in the outback, which I hadn't anticipated; the surprises it held in store really enriched my day-to-day research activities and prompted me to recognize how lucky I was to be pursuing this career.

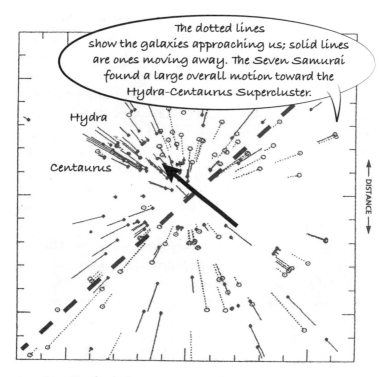

Figure 2.10 The dynamic map constructed by the Seven Samurai in 1986 (from Alan Dressler, *Voyage to the Great Attractor* [New York: Alfred A. Knopf, 1994]).

In Sydney, Professor Warrick Couch welcomed me to the physics department of the University of New South Wales. He'd agreed to let me work with him for a year. For me, a young student from France, Warrick exemplified the kind of professional path available to subjects of the Commonwealth. Born and raised in New Zealand, he'd done most of his studies in Durham, up in northeastern England, before heading to Australia to take a permanent position. His accent was hard to place: he didn't share the drawl and nasal pronunciation of most Australians, and his way of speaking was more clipped—maybe this was the New Zealand influence? At any rate, his deadpan humor and fondness for university culture and sports like rugby and cricket made him the perfect representative of the English-speaking world in my eyes. An outstanding astronomer, Warrick Couch is considered one of the most active members of the field in his country; he's devoted his career to studying the evolution of galaxies and the universe's expansion. He taught me a great deal about how to conduct observations. Now he directs the prestigious Australian Astronomical Observatory.

In Australia, over half of the population lives in one of the five main cities: Sydney, Melbourne, Brisbane, Perth, and Adelaide. Most of the physicists I met at the university lived "Australian style": an agreeable routine that involves enjoying meals out-doors, socializing at pubs almost every evening, and taking advantage of the fresh air—and water—at the seaside and urban oases. For my part, I had something else to do, too: conduct nighttime observations. I had the good fortune of being able to use a number of telescopes in New South Wales: the "Dish" at Parkes Observatory, which, at 64 meters in diameter, is excel-lent for observing spiral galaxies; the facilities near Narrabri, which is now home to the Australia Telescope Compact Array, a system of six dishes 22 meters in diameter also used for radio astronomy; and, to be sure, the instruments at the Siding Spring Observatory in Coonabarabran, where I spent much of the year, often conducting observations for weeks on end. All of these observatories are located several hundred kilometers from Syd-ney, to the west of the Blue Mountains, a small range running along the east coast. To make it to the telescopes, one needs to

Figure 2.11 View of the Siding Spring observatory in Warrumbungle National Park, New South Wales, Australia.

travel through the bush—a vast ecoregion some 800,000 square kilometers in size, mainly composed of eucalyptus forests and reddish earth.

These dusty roads offer few distractions. Every hundred kilometers or so, there's a small town, but on the whole, there are more birds—big white cockatoos with yellow crests and galahs (a kind of pinkish pigeon)—than the human beings with typical accents who inhabit these parts. My time in Coonabarabran left me with vivid memories. I felt like a real explorer: during the day I'd wander down the narrow paths of Warrumbungle National Park adjoining the observatory, looking for kangaroos, emus, and flocks of variegated parakeets, which I immortalized with my camera. Then, at night, I'd search for the Great Attractor with a much larger device: the UKST (United Kingdom Schmidt

Telescope). This instrument, which was built in 1973, has a mirror 1.8 meters in diameter and is designed to take in a large field of vision. This means that it collects light originating from a vast expanse of the celestial vault. A "photographic plate" 36 by 36 centimeters enables one to observe an angular area 6.5° by 6.5° (42 square degrees)—a feature unique to this type of telescope, which is called "Schmidt" in honor of its inventor. There are very few examples of this type of telescope in existence.

I participated in the rejuvenation of this ageing telescope by testing new technology that promised to save a great deal of time when conducting observations: a multifiber spectroscope named FLAIR (Fiber Linked Array Image Re-formatter). It takes several hours to record the spectrum of an elliptical galaxy in order to measure its distance. Back in France, at the early stages of my doctoral work, I had measured about a hundred elliptical galaxies one by one over the course of several months at the Haute-Provence Observatory. There, the spectrograph was mounted on a telescope with a mirror 1.93 meters in diameter—a bit larger than the one I was using in Australia. It took about two hours of data acquisition to obtain a spectrum we could use. Because the telescope in Coonabarabran had a smaller diameter, it took between four and five hours to obtain a spectrum of equivalent quality—that is, practically half a night of observation!

The advantage the new "multifiber" technology offered was the ability to record the spectra of some hundred galaxies simultaneously. In other words, it allowed researchers, who are always racing against the clock (and each other!), to save time. The process went like this. On a plate of glass presenting a photographic image of the skies (taken beforehand), I located a zone containing eighty elliptical galaxies. Then I glued tiny prisms—barely a couple of millimeters in diameter—to the spots where the galaxies were located. These prisms made it possible to redirect the light from a given galaxy down an optical fiber. The photographic plate, now fitted out with tiny prisms and tentacle-like fibers, resumed its previous position in the focal plane of the telescope, and the optical fibers were joined to the spectroscope (an instrument two meters long, located at the foot of the telescope) in the dome. At night, the light from the galaxies illuminated the prisms; the digital CCD (charge-coupled device) sensors that had replaced the old, silver photographic process equipment could get to work and begin measuring spectra.

Described in such summary fashion, the process sounds straightforward and simple enough. In fact, however, actually performing the task was not so easy. Sticking almost a hundred optical fibers, 0.1 millimeters in diameter and 11 meters long, to the precise spot of each galaxy proved quite time-consuming—especially for someone more used to typing at a computer keyboard than to holding tiny filaments still on a glass plate while waiting for ultraviolet-sensitive glue to dry! Preparing just one plate would take up almost a whole afternoon. All the while, I'd try to start the wiring on the next plate, just in case meteorological conditions made it possible to observe another part of the sky toward the end of the night.

Finally, when darkness fell, I'd set up the fully wired plate in the telescope and perform the adjustments necessary for gathering spectral data. Observation periods lasted from four to five hours; in more concrete terms, however, they were composed of sessions twenty to thirty minutes long, during which I had to make sure, among other things, that the telescope's movements were properly compensating for the earth's rotation, so that the celestial objects under investigation remained in focus. I also had to go out on the walkway around the dome tower to check for high-altitude clouds, which could compromise the whole observation period. I spent night after night climbing the stairs up to the telescope and back down again, always in a hurry to return to the ground-level optical lab where I was trying to finish wiring the next plate. These conditions made it pretty hard to find a spare moment for analyzing data! All the same, I had to do so, if only to know whether the recording quality was adequate and we could devote the following night to another part of the sky—or if, on the contrary, we'd have to keep our sights trained on the same area and leave the current plate and configuration of optical fibers in place. The first step of processing measurements—and the basis for any analysis at a later stage—is called "data reduction." This, then, was the first task I needed to perform, and it had to happen no matter what: when everything was running smoothly, but also when inclement weather forced us to close the dome; alternatively, I'd have to do it when I finally woke up late in the morning, after only a few hours' rest.

I can still remember the checklist I had to follow for reducing data from the spectrograph. First, corrections had to be made for major effects of atmospheric distortion on the grand scale. Then the overall image the sensors had provided needed to be "cut up" into individual spectral fields, one for each optical fiber. Average illuminance and wavelengths had to be calibrated, cosmic rays removed, and so on... Once I'd analyzed the raw data, I could tell which galaxies I should observe again the following night—or not. Then it started all over: gluing prisms to key points on the photographic plates and attaching optical fibers... The process could go on for three weeks in a row. Needless to say, the period of the full moon promised welcome rest. In actual fact, however, even during the less hurried pace of

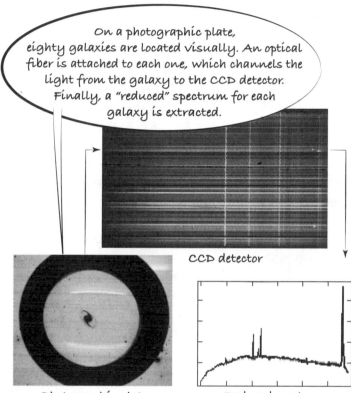

On a photographic plate, eighty galaxies are located visually. An optical fiber is attached to each one, which channels the light from the galaxy to the CCD detector. Finally, a "reduced" spectrum for each galaxy is extracted.

CCD detector

Photographic plate

Reduced spectrum

Figure 2.12 Obtaining spectral data with FLAIR.

what astronomers call "bright time," I had to sift through data to make my galactic maps. Circumstances permitting, I'd get out of the bush for a few days and head for Sydney.

Back to Earth

Unfortunately, this method of multifiber spectroscopy, which was meant to enable us to measure a large number of elliptical galaxies in as little time as possible, failed to produce the results desired. For one, all the fibers had the same diameter, which wasn't always the right size for the objects surveyed. A similar problem arose with respect to exposure time, which, with this system, was identical for all the galaxies. In some instances, the signal was adequate, but results proved unsatisfactory for less luminous galaxies. Above all, inasmuch as the galaxies were located at varying distances, absorption lines occurred at a large wavelength interval (on account of different Doppler shifts from one body to the next); in consequence, the overall spectral resolution employed was rather weak. Such low resolution prevented us from determining line widths with sufficient precision to be able to calculate the velocity dispersion of the stars in the galaxies—and that was the whole point of our observations!

And yet, despite the relative failure of the project, I hadn't wasted my time in Australia. When I recognized that precise measurements couldn't be obtained by means of the Faber-Jackson method, I quickly turned to using the redshift method, which I applied to more than 600 galaxies. In addition, the Parkes radio telescope (located farther to the south) enabled me to measure the distance of several spiral galaxies, which I calculated with the Tully-Fisher relation. At long last, by bringing together the measurements I'd made in France and Australia with those

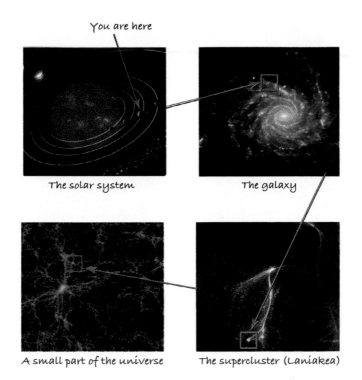

You are here

The solar system

The galaxy

A small part of the universe

The supercluster (Laniakea)

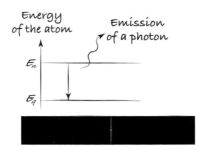

Energy of the atom

Emission of a photon

E_n

E_q

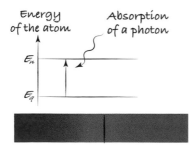

Energy of the atom

Absorption of a photon

E_n

E_q

Plate 1 (top) From the solar system to the observable universe: you are here!

Plate 2 (bottom) Spectral transitions in the course of a photon's emission and absorption.

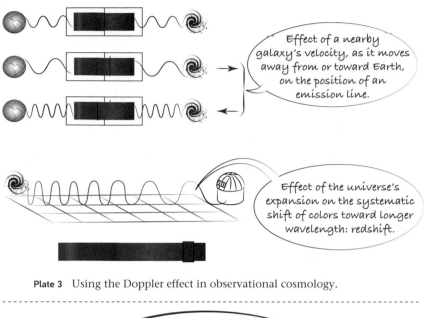

Effect of a nearby galaxy's velocity, as it moves away from or toward Earth, on the position of an emission line.

Effect of the universe's expansion on the systematic shift of colors toward longer wavelength: redshift.

Plate 3 Using the Doppler effect in observational cosmology.

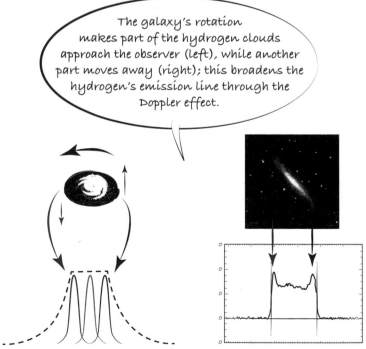

The galaxy's rotation makes part of the hydrogen clouds approach the observer (left), while another part moves away (right); this broadens the hydrogen's emission line through the Doppler effect.

Plate 5 Broadening of the hydrogen emission line from a spiral galaxy.

Plate 4 The main telescopes used to create our "modern" data sets. Left to right, top to bottom: Parkes (Australia), Green Bank (West Virginia), Mauna Kea (Hawaii), Spitzer (NASA), Hubble (NASA), Arecibo (Puerto Rico).

Plate 6 Spiral galaxies on parade. Left to right, top to bottom: NGC3521, NGC986, M33, NGC4565, NGC1232, NGC6118, NGC187, NGC253, NGC1365.

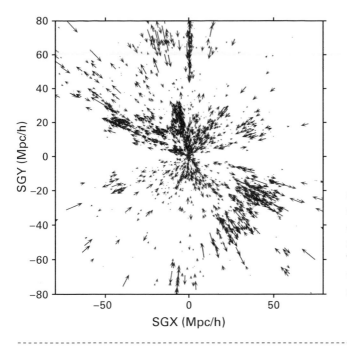

Plate 7
Our Cosmicflows-2 dynamic map of radial velocities. The arrows indicate the direction of movement.

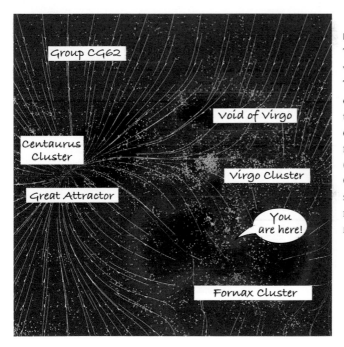

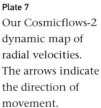

Plate 8
The purpose of visualization. This map has been constructed with the same data that our first dynamic maps presented (as in figure 3.8). Our visualization software now permits more detailed representation.

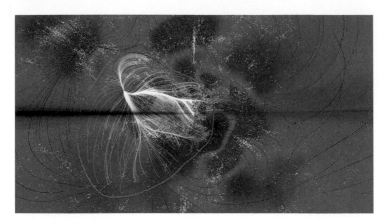

Plate 9 Evidence of our Laniakea Supercluster through a visualization from the CF2 dataset.

Plate 10 An image of Laniakea projected onto a building façade, Lyon Festival of Light, 2014.

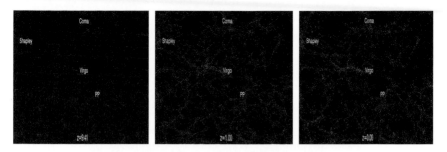

Plate 11 Time-lapse produced by a constrained simulation based on CF2. Here we can see the birth and development of our corner of the universe, from left to right at the ages of 500 million years, 6 billion, and 13.8 billion years.

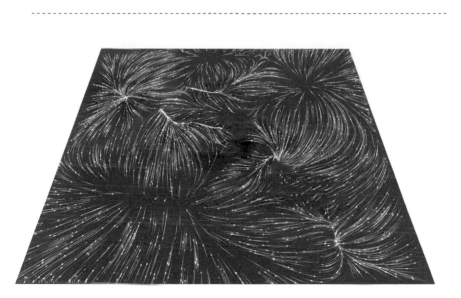

Plate 12 First map from the Cosmicflows-3 generation. Here we can see that the cosmic flows show the other watersheds surrounding Laniakea.

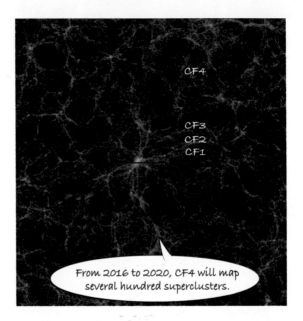

Plate 13 The sizes of the four Cosmicflows programs.

in the LEDA database, I managed to complete my PhD thesis in 1995 and construct a dynamic map comprising 1,376 galaxies— three times the number of galaxies the Seven Samurai had mapped in 1988.

But counter to the hope I'd secretly been nourishing, doing so didn't fill in all the blanks between large-scale structures. The map I now had before me didn't look like a cocoon with smooth, unbroken lines—at most, it looked like one that had been damaged at various points, which the butterfly had simply abandoned! For instance, I couldn't make out any galaxy fila-ments connecting two of the structures that were supposed to form the cocoon's surface: Perseus-Pisces and Pavo-Indus. If any-thing, the picture confirmed the presence of a void, which, given the "bubble" structure of the universe, was hardly surprising. But it was possible to see partitions forming between galaxies—hints of new bubbles in the Local Universe. Even more importantly, in my opinion, the dynamic map seemed to confirm the move-ment of local galaxies toward a region located in the direction of the Great Attractor, in keeping with the findings of the Seven Samurai. Interpretation put the Great Attractor at the edge of the map, at a distance so great it defied the instruments of the day. The time had not yet come to learn more about the nature of this mysterious object...

A Dead End

At the time, the first publications by Faber and her team were causing quite a stir; for a good decade, cartography enjoyed a considerable vogue. Groups of researchers in North America, Europe, and Australia competed to produce the biggest and most accurate maps, all of them hoping to identify the Great Attractor first. After all, what was at stake was more than an immense, mysterious mass. The Great Attractor might also provide undeniable proof of dark matter—and, as such, the passport to a Nobel Prize! Finally, in 1999, an international conference was held and the various teams presented their results. No one had made a breakthrough. Clearly, technological limitations were proving insurmountable: telescopes weren't powerful enough to capture the nature of the Great Attractor.

In consequence, many specialists in the field gave up on this particular area of research. Some of them, tired of being unable to conduct experiments that would validate or invalidate their hypotheses, stopped making observations and, instead, focused on computerized simulations, especially of dark matter, to explore how these large structures form. Who could blame them? Astrophysics is an odd field—and not least of all in terms of its methods. Broadly speaking, "classic" science always takes the same approach: following the observation of an unexpected result, a new line of questioning emerges, prompting the researcher to venture a hypothesis within the framework of an existing theory. He or she designs and conducts an experiment to test the hypothesis; the outcome will allow him or her to confirm the theory, modify it, or discard it. In contrast to practices in other fields, however, programs of experimentation in astrophysics are limited to observing the light emitted by stars. There aren't exotic collisions like the ones nuclear physicists devise; chemical elements don't get mixed together, and projects don't involve finding ways to conduct research on ever smaller scales… It's not hard to see how being unable to work with objects that are so large and so far away might prove discouraging for some.

But I didn't lose heart. I made the most of this setback to familiarize myself with how computerized simulations are used.

In order to do so, I spent a year and a half at the Max Planck Institute for Astronomy in Heidelberg, Germany. And in the course of attempting to reach a more systematic understanding of galactic structures on a grand scale, I developed a passion for trying to characterize large galaxy structures more systematically—in other words, a passion for fractals.

After this postdoc—a necessary rite of passage for researchers at the beginning of their careers—I had the good fortune to obtain a permanent position as *maître de conférences* (senior lecturer, or assistant professor) at Claude Bernard University in Lyon, France. I made the most of the fact that my life was now a bit steadier without the interruptions of extended observation campaigns. I embraced my new position and role, designing a new course of study in astrophysics for students at all levels, from the first to the final year. Eventually, I also became the mother of three children. Like many other women—and men, too—I faced the challenge of striking the right balance between family and work.

3 With Fresh Eyes

The observation method we used for measuring the peculiar motion of spiral galaxies by means of new, high-performance telescopes.

Everything's Accelerating!

Most researchers committed to using telescopes—including me—were at a loss when the dynamic cartography of the universe's large structures failed to go beyond the status quo. Fortunately, the questions we face when trying to penetrate the world's mysteries are hardly going to run dry. Indeed, the more our knowledge expands, the more questions arise! At the beginning of the 2000s, cosmic observers were gripped by the likely fact that the universe is in the course of accelerating its expansion. In 1998, two international teams had produced evidence of the phenomenon based on watching about fifty type Ia supernovas (SN Ia for short) located at a vast remove from us. In 2011, the discovery led to the Nobel Prize in Physics.

In broad terms, a supernova is a very large star in the process of exploding. There are two main types, one of which is the kind that interests us here: the SN Ia. The first bit of good news for cosmic cartographers is the fact that—at least in theory—Ia supernovas all produce about the same absolute luminosity as they burn up. This category of object, known as the "standard candle," proves extremely useful. Measuring a standard candle's apparent brightness—a level of luminosity that is well known— gives us the means to calculate the distance at issue. Thanks to redshift, which is determined in light of the Doppler effect, we can calculate the Hubble constant, H, that "matches" the star.

The second bit of good news concerns the fact that a single supernova is as bright as a whole galaxy. Therefore, we're able to detect them even at a great remove—billions and billions of light-years away. It follows that we can determine the Hubble constant for supernovas located at varying distances—that is, at different epochs in the past—which, in turn, allows us to test how the universe's expansion varies over time. In other words, it's thanks to supernovas that scientists can retrace the history of this expansion.

Doing so leads to something truly unexpected. When they applied Hubble's Law to the most remote supernovas, the method's pioneers recognized that these bodies were even farther away than their recessional velocities would seem to indicate. It looked like the galaxies were speeding up as they traveled! To explain the strange phenomenon, researchers postulated the existence of energy particular to empty space—that is, energy with different qualities than classical energy, an energy that tends to drive faster galaxies apart. In other words, they proposed the model of "dark energy"—a new arrival on the stage after dark matter! Needless to say, the universe harbors many mysteries...

The Value of the Hubble Constant

The acceleration of expansion is measured by comparing the rate of expansion as it was in the universe's past, which is estimated using distant supernovas, with its value in the universe today; the latter value, H_0, is measured using the nearest galaxies. It is of paramount importance, then, to determine H_0 as precisely as possible. Since Edwin Hubble conducted his pioneering studies, considerable research has been devoted to measuring H_0—the rate of cosmic expansion today; for a long time, the values assigned to it have oscillated between 50 and 100 km/s per megaparsec.

There are currently several teams working to measure this constant by extremely different experimental means. Measurements obtained using the pulsation of Cepheids are widely considered the most accurate. For over two decades now, astrophysicist Wendy Freedman and her team have counted as the world experts in this. One of their recent measurements was $H_0 = 72$ km/s per megaparsec. In other words, one megaparsec (3.26 million light-years) of space expands by 72 km each second. The value in question is being measured with increasing accuracy: in 2016, this accuracy stood at ± 2 km/s per megaparsec. On the basis of galaxies' peculiar velocities, my own team has produced a measurement of this rate of local expansion. We found quite the same value as Freedman, but our margin of error was much greater: $H_0 = 74 \pm 5$ km/s per megaparsec. A slightly different value of 69 km/s per megaparsec was determined based on recent cosmic microwave background readings taken by the Planck satellite operated by the European Space Agency. It is clear that the range of possible values has diminished significantly since the days when it was given as 50–100 km/s per megaparsec.

Like many of my colleagues at the time, I busily set out to hunt for supernovas. That said, using Ia supernovas as standard candles for calibrating Hubble's Law presents disadvantages, too. Research along these lines can prove fairly tedious inasmuch as few examples of type Ia supernovas are out there. The phenomenon occurs only in binary systems, when two stars are close enough to orbit each other. Although binary systems are common enough in the universe, for an SN Ia to admit observation, one of the stars must be almost dead (at the white dwarf stage), and the other must be a giant in the process of expanding. In such a case, the large star pours a stream of gas onto the smaller star which, when it reaches a certain critical mass, turns into a thermonuclear bomb! The phenomenon is so rare that the last SN Ia in our galaxy was observed in 1604; for three weeks, Kepler and his contemporaries could see it in full daylight, even though it was happening 20,000 light-years away. In brief, a Ia supernova is an infrequent event, an explosion that doesn't last for long.

The best way to detect supernovas is to photograph the same part of the sky, night after night, while keeping watch to see if some new, luminous object appears on the image. When this occurs, one has to act quickly and measure its apparent brightness and velocity in terms of the Doppler effect. Looking for a supernova is a bit like fishing or a cricket match: patience and quick reaction time are indispensable.

After spending several years making these observations, my colleagues and I published our findings, which confirmed that cosmic expansion is accelerating. Since then, our article, which appeared in 2006 in the *Astronomy & Astrophysics Review*, has been cited by 2,000 other scientific publications; it's become a key point of reference in the field...

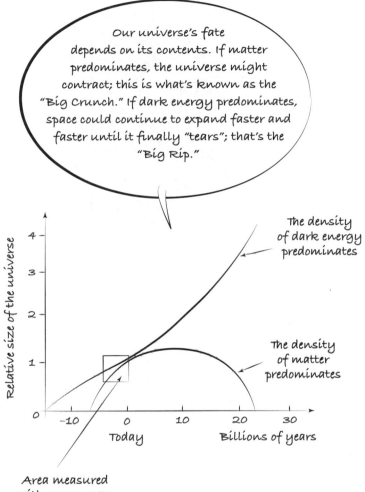

Figure 3.1   What is the fate of a universe in accelerated expansion?

Wendy Freedman

Figure 3.2

Born in 1957, Wendy Freedman is a Canadian-American researcher who completed her training in Toronto. At the age of 27, she joined the Carnegie Observatories in Pasadena, where she assumed the position of director in 2003. Today, she is a professor at the University of Chicago. Freedman also oversees the construction of the giant Magellan telescope at the Las Campanas Observatory in Chile. Throughout her career, she has focused on using Cepheid stars to measure the distance of galaxies other than our own. One of her many notable accomplishments is leading a team of thirty researchers working on the Hubble space telescope (Hubble Key Project). On the basis of their observations, Freedman and her colleagues first determined the value of the Hubble constant to lie at 72 km/s per megaparsec. When this figure (accurate to 10%) was published in 2001, it put an end to a heated debate running since the 1930s; the value assigned to this fundamental constant had varied between 50 and 100 m/s per megaparsec, depending on the research team. Freedman received the prestigious Gruber Cosmology Prize in 2009.

How the West Was Won

I arrived in Hawaii in June 2006. Having spent several years analyzing the outcome of calculations and trawling for supernovas
in the ocean of dark energy, I was eager to get back to hunting
for dark matter by studying the movements of galaxies in the
Local Universe. I was especially happy to do so in light of the
fact that, since 1999, some very promising advances had been
made in telescope technology. It was high time to return to the
quest for the Great Attractor! I discussed my plans with Professor Brent Tully, the inventor of the famous method bearing his
name that is used to measure the distances of spiral galaxies.
Right away, he invited me to work with him for a full year at one
of the most distinguished observatories in the world, the Institute for Astronomy (IfA) at the University of Hawaii. Thus began
a fruitful collaboration that continues to this day. Since then,
I have made many trips back to his institute—usually during the
summer, when teaching breaks at the University of Lyon give me
more freedom to pursue my research.

Australia, Hawaii…There are certainly worse places to work!
That said, astronomers didn't choose these places just because
they wanted to enjoy the sea and sand. Scientific criteria for
choosing a good location for an optical telescope—that is, a telescope that collects light from the visible spectrum—are a bit more
exacting than that. To begin with, the site needs to be located
far from human habitations and the light and particle pollution
they produce. It might seem surprising, then, that many European cities have an observatory. But these prestigious scientific
monuments were built at a time (from the seventeenth century
on) when manmade pollution was limited; today, they remain
important centers for research but are hardly ever used for
observing the sky. A good site should also be located somewhere

dry, which is often the case at high altitudes. In the mountains the atmosphere is thinner, which also minimizes the effects of turbulence and makes it easier to obtain focused images. The measure of clarity at a site is called the "seeing." It represents the average angular dispersion of a star's image when measured at different times in the course of a night. The moister and more agitated the atmosphere, the blurrier and more distorted images appear. The smaller the seeing is, the better the measurements will prove; without any atmosphere at all, the value is zero, which is why telescopes orbiting in space offer great advantages. Hawaii provides excellent conditions for observation; here, the seeing lies in the order of a half arc-second (an angle of $0.5/3,600°$).

Another important consideration is that telescopes need to be located in the northern and southern hemispheres in order to cover the entirety of the celestial vault. In the southern hemisphere, most observatories are found in Australia, South Africa, and Chile. In the northern hemisphere, France has a number of professional observatories, with radio telescopes at Nançay, on the Bure Plateau, and optical telescopes at the Haute-Provence Observatory, Pic du Midi, on the Calern Plateau, and elsewhere. These sites permit spectroscopic measurements even without optimal conditions for "seeing."

In Hawaii, optical telescopes have been installed at an altitude of 4,200 meters near the summit of the dormant volcano Mauna Kea ("White Mountain" in Hawaiian). It's technically the tallest mountain in the world because it's an enormous shield volcano resting on the ocean floor, 6 kilometers below sea level. As such, Mauna Kea measures 10 kilometers in height; not even Mount Everest can beat that! The Hawaiian archipelago formed over the course of millions of years; the oceanic plate from which the islands emerged slowly traveled northwest, passing above a "hotspot"—that is, a site where magma flows. On occasion, this magma broke through a thinner expanse of the plate and created a new island. The "big island" where Mauna Kea lies is the youngest such formation, located at the southeast of the archipelago. Being so high above sea level, the Hawaii Observatory is famous for its excellent views. Moreover, the closest coasts are thousands of kilometers away in every direction: California is 4,000 to the east, Japan 6,000 to the west, and Sydney 8,000 to the southwest! This means that overpopulated coastal regions don't spoil the darkness of the Hawaiian skies at night. The closest city of note is Honolulu, the state capital, where the university's institute of astronomy is located. About the size of Lyon, Honolulu is located on the island of Oahu, 300 kilometers from Mauna Kea—too far away to flood the telescope with unwelcome light!

What's more, the continents are so far away from Hawaii that they have no influence over the regional meteorology, which is perfect for making observations. During the day, water in the tropical forests evaporates in the heat; then, in the evening, it condenses near the peaks, where it rains on the windward side. As a result, Hawaii almost always has clear skies at night both at sea level—where tourists gape in admiration at the sunset and starry skies—and at very high altitudes above local condensation, where astronomers pursue their research. Its location on the Tropic of Cancer gives Hawaii a further advantage, too: from here, it's possible to survey the whole northern hemisphere as well as a large part of the southern hemisphere.

Finally, in 1959, Hawaii became the fiftieth state of the United States of America. The USA had been interested in the archipelago for over a century, initially for growing sugar cane and pineapples. Later, its strategic position in the middle of the Pacific made

Figure 3.3 The main telescopes used to create our "modern" data sets. Left to right, top to bottom: Parkes (Australia), Green Bank (West Virginia), Mauna Kea (Hawaii), Spitzer (NASA), Hubble (NASA), Arecibo (Puerto Rico). (Also plate 4.)

it an important point of departure for the navy and air force on expeditions to the Asian continent. In more recent years, the USA has established the necessary infrastructure to make the most of the archipelago's potential for hosting tourists from California and Japan. The observatory owes a great deal to investments made by NASA and assorted American and international partners. The process has not always gone smoothly. Wounded by the partition of their land at the end of the nineteenth century, native Hawaiians harbor resentment against Western civilization, and it can well up to the surface at any time. During the summer of 2015, a series of protests erupted against the installation of a new telescope, the TMT (Thirty Meter Telescope), just below the summit of Mauna Kea. Hawaiian tradition considers mountain peaks sacred sites. In consequence, no telescope has ever been, or will ever be, built at the very top of Mauna Kea. Astronomers held talks with the islands' inhabitants for seven years to reach an agreement. Negotiations included dismantling old telescopes before constructing the new one, and providing substantial resources for funding education across the archipelago. That said, opposition persists; it is unclear to me what the destiny of the TMT will be in Hawaii.

Spiral Galaxies In Motion

Though Canadian by birth, Brent Tully has been Hawaiian for over forty years. He lives in a small house on a bay surrounded by palm trees. When he's not conducting research, he enjoys spending time in his garden and gazing at the magnificent sight of waves breaking over the coral reef. It's his good fortune to be able to sit and think about the future of the cosmos with his feet in the water.

Tully enjoys worldwide renown in our field. In 1977, he and his colleague Rick Fisher discovered the relation now bearing their names, which revolutionized methods for determining extragalactic distances. For what it's worth, some of the credit also goes to France; he was a postdoctoral researcher at an institute in the southern part of the country when he carried out and published his work. What the Faber-Jackson method is for elliptical galaxies, the Tully-Fisher method represents for spiral galaxies. These two modes of calculating distances, which apply to different types of galaxy, were discovered at around the same time, and they complement each other—especially since their objects don't occupy the same parts of the universe. As a rule, spiral galaxies are found running the length of filaments of matter and away from the central knots of the cosmic web. The most interesting ones, which are also the most isolated, are called "field galaxies." They display the most pronounced motion—under the gravitational influence of larger clusters— and therefore prove ideal for tracking. In contrast, elliptical galaxies usually occupy the center of knots where peculiar motion is chaotic. "City galaxies" are less interesting than "field galaxies," when research concerns the coherent large-scale dynamics of the Local Universe. This consideration prompted me to leave my first love—elliptical galaxies—and change my focus to spiral galaxies. Doing so turned out to be a decisive step toward discovering Laniakea.

A major difference between the two methods involves the kind of telescope employed. For elliptical galaxies, we measure the broadening of absorption lines caused by the central stars' velocity dispersion. Since these lines fall within the visible spectrum, optical telescopes are used. Spiral galaxies, on the other hand, tend to present hydrogen clouds at the perimeter of the galactic disc. Occasionally, when at rest, the hydrogen emits a

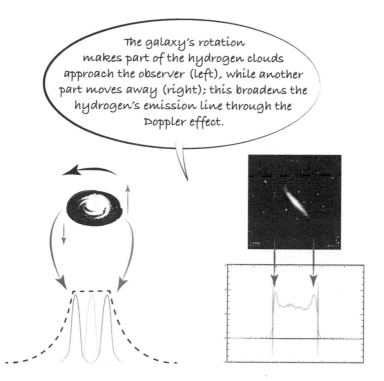

Figure 3.4 Broadening of the hydrogen emission line from a spiral galaxy. (Also plate 5.)

line in the radio spectrum at a frequency of 1.4 gigahertz (corresponding to a wavelength of 21 centimeters). Radio telescopes are used to observe this emission line.

To get a rough idea of the process at work, one can make a "spiral galaxy" at home with a small plastic pinwheel. Looking from the side as it spins, we can see certain blades (say, those on the right) moving away while those on the opposite side (the left, in this case) draw closer. In the same way, when it's possible to observe a spiral galaxy from the side—and not from the front—the hydrogen emission line appears larger because of the Doppler effect: at one edge of the galaxy, the hydrogen clouds are moving away from us (toward longer wavelengths), and at the other edge, the clouds are moving closer (toward shorter wavelengths). The rotation of the galaxy on itself gives the line an easily identifiable and measurable M shape.

This broadening of the spectral line depends directly on the galaxy's rotational velocity. But just as planets orbit the sun because of its gravitational attraction, the bigger a spiral galaxy is, the faster its stars and gases revolve around its center. Accordingly, when we know the rotational velocity of a galaxy, we can estimate its mass, which allows us to deduce its total absolute luminosity, a value that stands around 10^{37} watts. By comparing this value to the apparent brightness measured with optical telescopes, we can then determine the galaxy's distance. The Tully-Fisher method is illustrated in figure 3.5. In 1977, these two North American astronomers worked out the relation using just a dozen spiral galaxies. At the time, the distances were calculated with about a 30% margin of error. Since then, much larger samples have been tested, and the degree of accuracy is constantly improving. In the best cases, we can now measure the distance of spiral galaxies with an 8% margin of error.

It's not hard to understand how cosmographers are obsessed with accuracy. During my year in Hawaii, I wanted to perfect the Tully-Fisher method by focusing on the collection and analysis of data. I managed to come up with several improvements. For example, we knew that taking the morphology of each galaxy into account would help us determine the apparent brightness of spiral galaxies with greater precision. To this end, we employed a technique known as surface photometry. The process involves accumulating luminous fluxes produced by the "surfaces" of the galaxy on an extended scale, starting with the central bulge, and

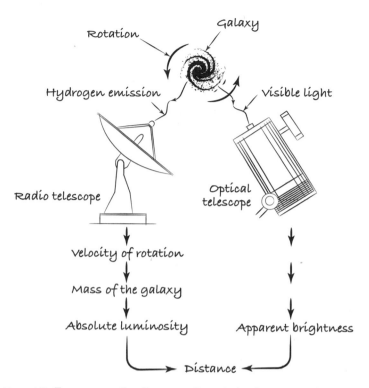

Figure 3.5 To measure the distance of a spiral galaxy using the Tully-Fisher method, we need to observe using two types of telescope: a radio telescope and an optical telescope.

paying particular attention to peripheral regions. This method is much more sophisticated than the single, rougher measurement previously employed, which is known as aperture photometry, but it requires an excellent seeing; needless to say, we made the most of Mauna Kea's ideal location.

But I didn't need to use the optical telescopes at this exceptional site just for surface photometry. Measuring hydrogen lines with a radio telescope is very expensive. To make the most of the

time we were given to use the radio telescope, we had to select
the best candidates for observation beforehand.

Beauty Contest: Miss Spiral

How does one tell whether a galaxy merits observation through
a radio telescope? First of all, it has to be a spiral galaxy in order
to contain hydrogen gas.

Second, because we are looking for large-scale coherent
motions, and not for nearby chaotic interactions, the galaxy must
be located at some distance from its neighbors. An optical telescope is required for determining whether a given spiral galaxy
is sufficiently isolated, and it needs to be distinct both in terms
of its location on the celestial vault and in terms of depth. By
means of Hubble's Law, the galaxy's approximate distance is calculated on the basis of its recessional velocity. To this end, the
galaxy's redshift (already measured by other astronomers and in
our databases) is employed. This redshift also proves valuable
when radio measurements are made, because it allows us to
select the best wavelength window for gauging the broadening
of the hydrogen line.

The final criterion for picking a good candidate is also evaluated using optical observation: the disk of the galaxy should
be inclined relative to the celestial vault. A galaxy that can be
viewed "face-on" has a 0° angle of inclination; one that is visible "edge-on" has a 90° angle of inclination. As noted in the
pinwheel analogy, when a galaxy faces us directly, we cannot
detect line broadening by means of the Doppler effect. Spectral
line broadening depends on the galaxy's inclination. It is still
quite difficult to gauge this angle precisely; indeed, galaxies'
inclinations represent the greatest source of uncertainty in our

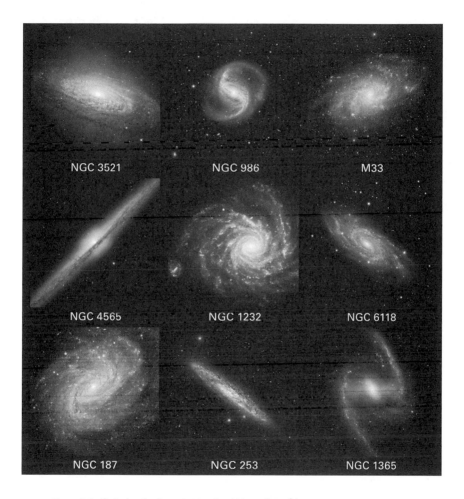

Figure 3.6   Spiral galaxies on parade. (Also plate 6.)

method. The more spiral galaxies are inclined—that is, the more we can observe them edge-on—the more precise our measurements will be. One of my closest colleagues, Igor Karachentsev, works on the largest Russian telescope, which is located in the Zelenchuk valley in the Greater Caucasus mountains. For the most part, his research focuses on galaxies with an extremely

Elliptical Galaxies versus Spiral Galaxies

Hubble and his galaxy-hunting colleagues quickly realized that galaxies come in several shapes: spiral, barred or unbarred, elliptical, lenticular, irregular ... More recently, scientists have also discovered a multitude of dwarf galaxies. When researchers want to determine peculiar velocity, the big galaxies—spiral and elliptical—provide the main points of reference. Because spiral galaxies contain a large amount of hydrogen gas, especially at the edge of their disc, the Tully-Fisher method is used. This method doesn't work for elliptical galaxies, which have all but exhausted their supply of cold hydrogen (which is a basic element in stars).

The current state of research suggests that elliptical galaxies are the result of collisions between spiral galaxies. These collisions produced shock waves that transformed the clouds of hydrogen into "nurseries" of stars. If we point a telescope set to the wavelength of 21 cm toward an elliptical galaxy, we see very few radio waves. What's more, inasmuch as elliptical galaxies don't display much internal rotation, we don't observe the spectral line broadening evident in spiral galaxies (which yields a characteristic M shape and facilitates very precise measurement of their rotational velocity). Where elliptical galaxies are concerned, we must be content with less pronounced broadening, which occurs only on the basis of the stars' motion in all directions (Faber-Jackson method). This means that results are much less precise.

(continued)

A further difference between the two types of galaxy—and one of the reasons we prefer spiral galaxies—is the fact that elliptical galaxies tend to be found in large clusters, at points where numerous galaxies' trajectories can intersect. Because this environment is much more crowded, their motion is chaotic and much less representative of large-scale cosmic flows than that of spiral galaxies, which are more distinct and form part of the "traffic" in space; we call these isolated spiral galaxies "field galaxies," because they allow us to plot the velocity field as a whole.

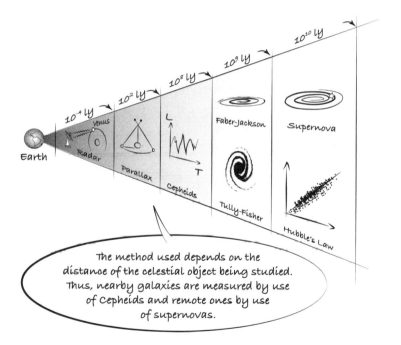

Figure 3.7 Different methods for estimating the distances of celestial objects.

high degree of inclination. Such galaxies are few in number, and my own studies included examples with a broader range of variation, anywhere between 45° and 90°.

A Job with Benefits

I went to Hawaii to conduct observations by means of a modest-sized telescope (one with a mirror 2.2 meters in diameter) and to code new algorithms in order to modernize the Tully-Fisher method. But I won't lie: I made the most of my time off to enjoy this small heaven on earth. It started with the leis offered in welcome as I got off the plane and continued with hula dances, magnificent beaches (some of them crowded, and others with hardly any people at all), abundant exotic fruit—from sweet mangoes and tangy papayas served with lime to guavas, coconuts, and lilikoi—and swimming in the company of tortoises, dolphins, and whales (while making sure to avoid sharks and moray eels).

But despite such splendors, I spent some of the most enjoyable time in my whole career carrying out nighttime observations. Wherever I may happen to be, I love strolling on the catwalk around the dome with no immediate purpose—listening to the sounds of cicadas at Saint-Michel-l'Observatoire, or watching kangaroos in Coonabarabran. It's an incomparable sensation, exultant yet calm, to feel so alone, up at the top of an icy mountain, and, at the same time, to sense that one is part of the vast universe—just by looking up at the starry sky.

Alas, the lightning pace of technological innovation, especially in the domain of communication, means that tomorrow's astronomers will have fewer opportunities to savor moments like these. I experienced the change firsthand during my time in

Hawaii. Needless to say, there was always a technician manning the dome, but I wound up spending most of the actual observation period supervising the night's overall progress from a control room at the University of Hawaii in Honolulu—an island located 300 kilometers away from the telescope on Mauna Kea! It broke my heart.

Something New from the East, Too

Brent Tully and Richard Fisher met as students and soon began working together on measuring galaxies. Professor Fisher has long suffered from a degenerative disease that, since the final stage of his studies, has prevented him from traveling to conferences. Paralyzed on one side of his body, he also has difficulty participating in oral exchanges. All the same, he's the master of radio astronomy data in the Tully-Fisher team. When I went to the Green Bank Observatory in West Virginia, which is home to the world's largest steerable radio telescope, I learned a great deal from Fisher; in large part, he's the reason for my expertise in the field.

As noted, technological advances have been proceeding briskly. This has certainly been the case for radio telescopes, including the one at Green Bank. In 1988, the predecessor of the current telescope fell apart; over the course of almost ten years, its gigantic replacement was built: a new generation of instrument equipped with state-of-the-art, "multibeam" technology. On the outside, a radio telescope presents an immense metallic dish. Radio dishes can be built on a massive scale because their "mirror" is significantly lighter than optical mirrors of the same size. Moreover, radio mirror surfaces don't need to be as smooth as optical mirror surfaces. In the field of astronomy, it's essential to minimize the diffraction of light waves, which causes the signal quality to deteriorate. This phenomenon occurs when irregular surfaces belong to the same order of magnitude as the wavelength being observed. The longer the wavelength, then, the greater the range of admissible imperfections. This is why optical mirrors must be polished until their surfaces are extremely smooth, with defects smaller than a hundred nanometers (that is, one ten-thousandth of a millimeter). The task proves impossible for mirrors more than a few meters in diameter. But radio telescopes work with longer wavelengths (ranging from a millimeter to a few meters), which makes larger defects (measured in millimeters) acceptable. In consequence, we can build metallic mirrors that are much larger than optical glass mirrors—up to hundreds of meters across. This does not entail a loss of quality, since a given telescope's power is measured mainly by the size of its light collector: a larger surface area means that more photons are collected.

The antenna of the Green Bank Telescope (GBT) is 110 meters in diameter, the length of a soccer field. It comprises more than 2,000 panels, which can be individually adjusted through as many tiny motors: planar deformation lies at a threshold of about 80 micrometers! A similar radio telescope exists in Effelsberg, Germany, which is a little smaller—100 meters in diameter. There are also two instruments that are larger, but immobile: the Arecibo telescope (300 meters), at the bottom of a meteor impact crater in Puerto Rico, and the FAST telescope (500 meters) currently being tested in China. To be sure, the Arecibo telescope can reach more distant galaxies than the GBT. What's more, even if its "lobe size" (the field of vision in radio astronomy) is smaller, this often represents an advantage inasmuch as it permits observers to measure the spectra of nearby galaxies, which would overlap if the beam were larger. But because it cannot move, Arecibo only permits one to observe 30% of the sky. In contrast, the GBT can pivot and take in 85% of the sky. When the dish is set at a vertical angle, for instance, it's possible to observe a target 5° above the horizon.

"Viewing Requests"

Its many advantages make the GBT a highly coveted instrument. In fact, less than a hundred years have passed since Karl Jansky, an American engineer working for the Bell Telephone Laboratories, accidentally discovered a source of radio interference located at the center of our galaxy. At the time, these extraterrestrial radio waves didn't garner much notice. Now, however, there are many astronomers who want to catch the radio signals, which are full of information that visible light does not reveal. Subsequent research has shown that what Jansky found was the throbbing emission of a supermassive black hole, the beating heart of our Milky Way. Radio waves are also emitted by pulsars, which are the compact remains of exploded supernovas, and, as we've already seen, from galaxies farther off in the universe...

The pressure factor for the GBT—the ratio of the number of observation hours requested by astronomers to the number of hours available in a six-month period—stands at more than ten, which is almost as much as the figure for the Hubble Space Telescope. In other words, 90% of requests submitted prove unsuccessful... The fact that most telescopes are available to the international community doesn't mean they are available at any given time! Photons can't be collected like butterflies: there are few "nets," and everyone wants to use them. First, researchers need to write up a research proposal describing the projected scientific impact and number of telescope hours required; one must also demonstrate the ability to use the instrument to its full technical capacity. Since requests are accepted every six months—once for observation during the summer, and once during the winter—an unsuccessful application can delay projects by a year! Competition for this "passport to heaven" is intense. Sitting on a committee to decide how time on a telescope is allocated is incredibly frustrating. One hears any number of excellent, well-planned ideas that are explained in detail and would be possible to perform quickly, but there just aren't enough telescopes to go around.

Needless to say, managing to obtain observation time is key in our race to understand the extragalactic world. For instance, in my own research, the procedure used to obtain a 21-centimeter hydrogen emission line spectrum requires several bunches of ten-minute exposure times. After making a first ten-minute exposure, the data is immediately reduced in order to determine whether the integration time has, in fact, been adequate, or whether further data must be collected. In the case of the most remote galaxies, it sometimes takes fifteen exposures. Luckily, measurements of this kind are much less demanding in terms of climatic conditions than most of my colleagues' projects. I can observe galaxies in radio day and night, whether the weather's good or bad. Consequently, I had the good fortune of seeing my research program categorized as "large" by the GBT time allocation committee—a qualification that really means "filler program." In practical terms, this meant that my collaborators and I could conduct our observations throughout the year, taking slots at short notice whenever meteorological conditions prevented other researchers from conducting their own measurements.

As such, days of heavy precipitation always brought intense periods of observation. My team managed to gather data even with a meter of snow on the telescope, trying to make it melt from the sun all day before handing back the telescope at night to our colleagues. During the day, we targeted ideal candidates: isolated, inclined spiral galaxies that would appear in projection close to the sun (but not *too* close, since the sun would interfere with the radio waves we wanted to record). We also picked galaxies at a relatively low point on the horizon; this allowed us to position the huge dish (110 meters in diameter, and 8,500 metric tons) so it faced the sun almost at a vertical angle—which, of course, helped melt and clear the snow. We call this a "natural dump." This arrangement saved a lot of time and labor, and spared workmen the task of climbing up 150 meters to sweep off the dish.

Recognizing One's Errors

Parallel to observations at Mauna Kea and Green Bank (which helped to expand extragalactic maps and fill out blank spaces where few measurements had previously been available), I started standardizing data obtained by other researchers. As noted above, observation campaigns are time-consuming; it's much more efficient to pool the work of different teams than to carry out a few new measurements one by one. I availed myself of findings in the public domain—in particular, results obtained with the Nançay radio telescope by a French team (which included my dissertation advisor, Georges Paturel) and data from the Arecibo telescope collected by researchers at Cornell University. Using these significantly larger data sets (more than 15,000 galaxy spectra), we succeeded in improving the accuracy of calculations using the Tully-Fisher relation. This involved replacing the old way of measuring the 21-centimeter emission line at 20% of the flux's maximal value with a more robust measurement at 50% of the average flux. The new approach freed us from errors inherent to the variety of emission line forms corresponding to the variety of spiral galaxy morphologies (barred, unbarred, grand design, loose arms, etc.).

This modification of the Tully-Fisher relation, combined with better surface photometry, amounted to a real improvement. Indeed, it's essential for researchers to make sure that results are solid at each step to prevent errors—which have a cumulative effect—from being too large in the end. Another example of adjustments we're still making is the following. To determine a spiral galaxy's apparent inclination in the sky, we calculate the ratio between the length of its minor axis and that of its major axis. These values can be evaluated by eye or with an algorithm, based on a photograph of visible light in the galaxy. Estimating a galaxy's dimensions is fraught. An estimate of inclination that's even a little off—say, by 10%—will entail a series of erroneous results. Our calculation of rotational velocity will be wrong, and so will our assessment of the galaxy's absolute magnitude. And our error in these two values, which we use in calculating the galaxy's distance, means that our final measurement of this distance will be off by 30%! Astrophysics offers many other examples of errors being compounded in this way. It's not hard to appreciate the extreme care that must be observed in the field and when interpreting results. As a rule, basic research benefits from healthy doubt; indeed, it's the motor that drives a rigorous course of experimentation.

Observational Artifacts

An observational artifact is a phenomenon artificially created by the conditions of the experiment itself.

One of the first artifacts that cosmologists are likely to encounter is known as the Malmquist bias. It's easy to understand how the difficulty of detecting a celestial body proves greater the farther its distance and the lower its apparent brightness are. At a

(continued)

small remove, all objects are readily visible, including less luminous ones; when distances are great, only the brightest objects can
be made out. Cosmologists work in full awareness that their cata-
logs of galaxies grow more incomplete as distance increases.

A careful look at extragalactic maps (for instance, my "Cocoon"
at the end of the first chapter) reveals long, thin structures systematically pointing in the exact direction of the observer's position.
In the 1970s, Brent Tully dubbed these features "fingers of God,"
a sweet joke since they are pointing straight at us. Actually, they
are artifacts appearing when galactic distances are being calculated
with Hubble's Law (on redshift maps, that is). Galaxies inside clusters have a turbulent velocity of their own, which adds up to their
average recessional (expansion) velocity. Thus, instead of appearing spherical, clusters look elongated along the line of sight.

These two examples plainly illustrate that artifacts present
dangers that cosmic explorers need to watch out for—lest they
reach erroneous conclusions and fall victim to ignoble, anthropo-
centric prejudices! Fortunately, our method of exploration man-
ages to steer clear of such pitfalls.

Our First "Modern" Maps

In 2008, thanks to meticulous observation and analysis, as well
as research conducted by other teams, we finally had a reliable
data bank with the positions and radial velocities of 1,800 galaxies. Aptly enough, it's called EDD, or *Extragalactic Distance
Database*. This collection is almost five times bigger than the
information contained in the survey that revealed the mystery
of the Great Attractor. The 1,800 galaxies in question occupy an
expanse that extends for about 130 million light-years around
our planet. To ensure that the entire scientific community could

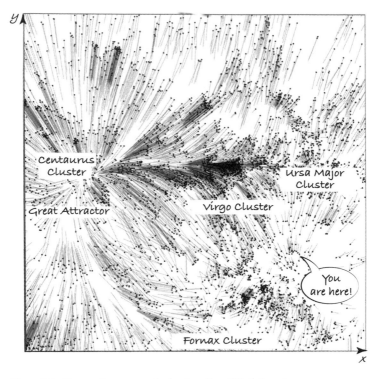

Figure 3.8 Dynamic map created using the CF1 data set.

benefit from our findings, we quickly published a series of cosmic maps showing the motions of these 1,800 galaxies—along
with information about 30,000 others for which we only had the
positions (determined through redshift).

In this context, let's look at one of the dynamic maps we constructed, measuring 260 million light-years across (figure 3.8).
Each black dot is a galaxy whose velocity is represented by a
gray line.

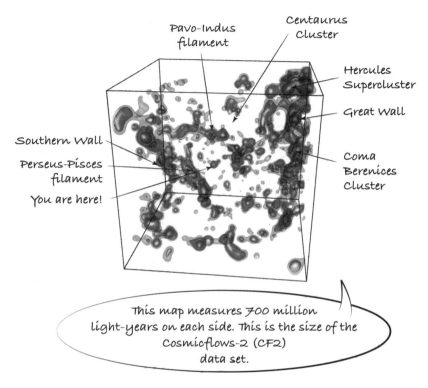

Pavo-Indus filament

Centaurus Cluster

Hercules Supercluster

Great Wall

Southern Wall

Perseus-Pisces filament

You are here!

Coma Berenices Cluster

This map measures 700 million light-years on each side. This is the size of the Cosmicflows-2 (CF2) data set.

Figure 3.9 A representation of large-scale structures.

Our Local Group is subject to gravitational attraction from the imposing Virgo Cluster, which contains 1,300 galaxies. In turn, the Virgo Cluster is drawn by a cosmic current toward the Centaurus Cluster. Significantly, the Centaurus Cluster is located in the area presumed to house the Great Attractor, too; evidently, it's where several galaxy currents converge, linking it to other clusters. We didn't know it at the time, but this map already showed what, a few years later, we would recognize as the skeleton of Laniakea. The Great Attractor hovers just at the edge of our dynamic map, veiled in mystery.

Using the 30,000 galaxies we know only by position (from velocities, assuming uniform cosmic expansion), we can map an even greater expanse, with a radius of about 350 million light-years (figure 3.9). In order to avoid observational artifacts (see "Observational Artifacts," above)—especially those that result from incomplete data (i.e., the Malmquist bias)—we decided to soften the edges of the density field, which minimizes the conspicuousness of the Local Region and puts large structures into relief. This map includes all the important large structures mentioned up to this point: the Perseus-Pisces Supercluster, the Pavo-Indus Supercluster, the Great Wall, the Southern Wall...

One can make out points of connection between the Southern Wall and the Perseus-Pisces Supercluster, as well as between the Hercules Supercluster and the Great Wall. Unfortunately, we don't have any information about galaxy velocities at this scale. Essentially, the representation we can give to such a vast volume yields a static map; these faraway galaxies have been measured only by means of redshift. The lack of dynamic data greatly restricts our understanding of the structures. We can identify connections statistically, but it's impossible to delineate, with any real precision, the large groups inside which galaxies follow consistent patterns of movement.

Cosmic Flows

All the same, I'm satisfied with the quality of these observations. Indeed, I'm now convinced that we made up for the technological delay that had kept us from seeing beyond the Great Attractor ten years earlier. The measurements I obtained during my year in Hawaii strengthened my belief that it is possible—imperative, even—to expand our map in all directions! To this end, we'd need to conduct research in both of Earth's hemispheres and make the most of every time zone, so daylight wouldn't ever slow us down... It would take the biggest telescopes, located at the best sites across the globe, because we needed to observe the most distant galaxies. Because they're so far away, they seem tiny and are harder to measure. In the northern hemisphere, we'd work with the telescopes in Hawaii, Green Bank, and Arecibo; for the southern sky, we'd make observations from Parkes (Australia) and Chile. Finally, we'd use the Hubble Space Telescope.

I recognized I'd need help to complete such an ambitious undertaking; conducting the measurements necessary for determining peculiar velocities takes time. Accordingly, I assembled a team of Europeans, Americans, Russians, and Australians—researchers including Nicolas Bonhomme, Dmitry Makarov, Sofia Mitronova, Maximilian Zavodny, and Baerbel Koribalski... The plan called for what's known as "open collaboration": members join the project at their own initiative and contribute what resources available to them permit. The project was called Cosmicflows, in keeping with the research focus. We named the first sample of 1,800 galaxies in motion Cosmicflows-1 (CF1). Immediately after, we launched a series of observations for Cosmicflows-2 to determine the velocities of 8,000 galaxies spread out over an expanse twenty times larger.

For several years, this team of men and women worked tirelessly on the CF2 project. It proved exhausting—both for my colleagues and for me. The most grueling days of an observation campaign would go something like this. At my house in the Alpes region of France, I'd wake up at 1:30 in the morning to connect with Green Bank (eastern United States). Until breakfast, I'd conduct observations, then go to sleep around 7:00. Three hours later, at 10:00, I'd get up again and connect to Hawaii to speak with colleagues and analyze the data being gathered there (in a time zone twelve hours away from France). At 2:00 pm, it would be time to teach at the university; then I would go to work at the IPNL (Institut de Physique Nucléaire de Lyon), where my research team is housed. After that, I could go home and have dinner with my family—before logging on at 10:00 pm to discuss the galaxies to target with Australian colleagues. Finally, I'd head to bed as early as possible, just in case we'd have to start everything from scratch tomorrow…That's the story of how, in 2009, I managed to carry out 480 nights of observation for Cosmicflows-2 by working in three time zones at once. It was worth it.

4 Reconstruction Games

The methods of analysis used to construct the detailed dynamic maps that enabled us to identify Laniakea as a galactic watershed.

Everything's Getting Bigger!

After four more years of gathering telescope data, we had Cosmic-flows-2 ready. It's much bigger than its predecessor: 8,000 galaxies with a known radial velocity, instead of 1,800. This sample is twenty times larger than the one used by the Seven Samurai, and the galaxies are distributed over a volume with a radius of 350 million light-years (as opposed to the 130 million light-years observed for Cosmicflows-1); to a great extent, it also covers the region where the Great Attractor is thought to be, 250 million light-years away. We'd established the largest database of galactic distances to date, with the most precise measurements ever.

At this juncture, we realized it was time to expand our research team. Up to this point, most members were cosmographers whose skills involved observation and analyzing raw data. We needed to add people with other areas of expertise, especially astrophysicists and data visualization specialists.

The Charms of the Wiener Filter

My first encounter with Yehuda Hoffman was more disconcerting than productive for me. If only in terms of appearance, we made a strange pair: Yehuda is about 30 centimeters taller than I am, and twice my weight—maybe more if you count the beard! But most of all, we came from two very different branches of astrophysics. He focuses on theory, whereas I concentrate on experimentation. As often occurs in situations like this, we found ourselves on completely different wavelengths—quite the irony for cosmologists! On top of it all, each of us had a few preconceived notions about the other's work.

A professor at the Hebrew University in Jerusalem, Hoffman had already been investigating large galaxy structures by means of a "reconstruction algorithm" based on peculiar velocities. Starting in the early 1990s, he'd used it to try to pierce the mystery of the Great Attractor, specifically by reconstructing the structures hidden behind the large strip of sky concealed by the galactic disc of the Milky Way; the brightness of its stars overwhelms us, and the dust it scatters blocks any extragalactic light that might be heading our way. The Great Attractor has evidently decided not to cooperate with earthlings: it seems to be located where it's difficult for us to observe it. Unfortunately, Hoffman's reconstructions—and the observations cosmographers had made—provided no clear answers. Thus, like many other astrophysicists at the end of the 1990s, he'd left this field of research. A decade later, when we met, his opinion hadn't really changed: he thought it was pointless to try to understand structures we couldn't observe even under good conditions.

For my part, I thought that his algorithm—which was about fifteen years old by this point—didn't have much to offer in terms of the data analysis we needed. The software he'd devised used a Wiener filter (named after the father of cybernetics Norbert Wiener, a mathematician, philosopher, and humanist). Video games are often designed with this type of filter. For instance, when programmers want a figure to move "on the cheap," they use animation based on just a few parts of its "body," say, the torso and limbs. The Wiener filter is suited for representing a large amount of visual data when few measurements are available. The *sine qua non* for the filter to yield good results is a few points of reference that always stick together; as a rule, this is the case for figures in video games, which are supposed to keep their bodies intact for as long as possible! The same principle holds for galaxies; here, gravitational interaction ensures "cohesion," provided the galaxies aren't too close or too far apart.

In this light, it's clear why Hoffman wanted to use an algorithm like this to reconstruct the motion of galaxies that are difficult to observe. I could appreciate other benefits, too. The algorithm enabled us to double the distances we could model in relation to observed distances; the results of our first tests—using the smaller Cosmicflows-1 sample (with a volume extending in a 130 million light-year radius)—already proved interesting. However, despite our best efforts, we still had to work with the few galaxies whose peculiar radial velocity had been clearly determined, in relation to neighboring galaxies whose position we knew only through redshift calculations (figures that stood at 1,800 and 30,000 for CF1, and 8,000 and 100,000 for CF2). If we used the algorithm effectively, we should be able to model galactic velocities by means of interpolation and devise what's known as a velocity field.

In the course of our discussions, Yehuda and I quickly recognized the benefits that a collaboration would yield. I managed to persuade him that our updated findings were precise, and he assured me of the reliability of the models his algorithm produced. We agreed that it was important to devote great care to obtaining precise measurements at every step of observation, and also that the software should be modified to perform the tasks of this study, in particular. Yehuda threw himself into the task of bringing the algorithm up to date and initiated Timur Doumler, a student from Germany working on his doctorate in Lyon, into the secrets of his filter's design. The team was growing.

Using this algorithm proved to be the second, major step toward discovering Laniakea; the first one had been my decision to observe spiral galaxies. Today, the significance of Yehuda Hoffman's contribution is clearer than ever; it's a real pillar of the project as a whole. Needless to say, rigorous observation and systematic treatment of raw data were crucial to our success, and the scheme of visualization enabled us to interpret and communicate the overall findings. However, there's no denying that a key element was using the reconstruction software to process peculiar radial velocities. Before we did so, our analysis of galactic motion had proved rather basic. Our "dynamic" maps incorporated the peculiar radial velocities of too few galaxies, which were represented by lines of various lengths (according to velocity) and following a simple color code (blue for movement in our direction and red for motion away from us); by this means, we'd tried to locate regions of high attraction.

Still today, Hoffman's powerful filter represents an integral part of our ongoing work on Cosmicflows-3, transforming the slightest measurement of radial velocity into a coherent pattern of movement—and thereby advancing our knowledge of large cosmic structures.

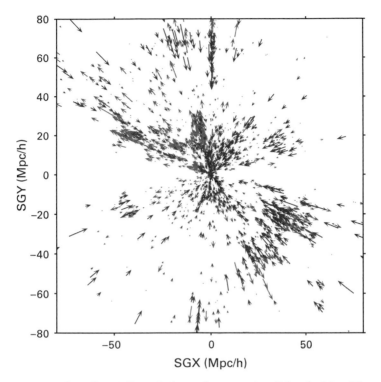

Figure 4.1 Our Cosmicflows-2 dynamic map of radial velocities. The arrows indicate the direction of movement. (Also plate 7.)

Let's Get Technical

Over the course of the three years it took to earn his doctorate, Timur Doumler wrote countless thousand lines of code so this sophisticated program would be able to complete the models desired. In the following, I'll try to explain how the filter operates without using complicated mathematical equations (for a few examples of which, see figure 4.2).

$$\left\{ u_i^o \right\}_{i=1,\dots,N}$$

$$u_i^o = \mathbf{v}(\mathbf{r}_i) \cdot \hat{\mathbf{r}}_i + \epsilon_i \equiv u_i + \epsilon_i$$

$$\left\langle \epsilon_i \epsilon_j \right\rangle = \left(\sigma_i^2 + \sigma_*^2 \right) \delta_{ij}$$

$$\mathbf{v}^{\mathrm{WF}}(\mathbf{r}) = \left\langle \mathbf{v}(\mathbf{r}) u_i^o \right\rangle \left\langle u_i^o u_j^o \right\rangle^{-1} u_j^o$$

$$\mathbf{v}^{\mathrm{CR}}(\mathbf{r}) = \tilde{\mathbf{v}}(\mathbf{r}) + \left\langle \mathbf{v}(\mathbf{r}) u_i^o \right\rangle \left\langle u_i^o u_j^o \right\rangle^{-1} \left(u_j^o - \tilde{u}^o{}_j \right)$$

$$R_{ij} \equiv \left\langle u_i^o u_j^o \right\rangle = \left\langle u_i u_j \right\rangle + \left\langle \epsilon_i \epsilon_j \right\rangle = \hat{\mathbf{r}}_i \left\langle \mathbf{v}(\mathbf{r}_i) \mathbf{v}(\mathbf{r}_j) \right\rangle \hat{\mathbf{r}}_j + \left(\sigma_i^2 + \sigma_*^2 \right) \delta_{ij}$$

$$\chi^2 = \frac{1}{N} \frac{u_j^o R_{ij}^{-1} u_i^o}{2}$$

Figure 4.2 Equations from Yehuda Hoffman's velocity reconstruction algorithm.

The Wiener filter is a conservative, Bayes-type estimator. In the simplest terms, it's meant to find the balance between observational data and a theoretical model. In zones where we know many peculiar radial velocities, the filter will tend to use observational data to calculate the velocity field; conversely, for areas where we haven't obtained many measurements—for instance, in the Zone of Avoidance or at the edges of the region surveyed—the filter will opt for the initial theoretical model, which must be carefully chosen. The one we use is called the Lambda Cold Dark Matter model (ΛCDM for short), which best describes the development of the universe as we understand it today. Λ (the Greek letter lambda) represents the cosmological constant for the universe's expansion, and cold dark matter is the dense mass responsible for the large majority of the gravitational attraction in a particular region of interest. The model's parameters are being improved constantly, so the theory agrees with observations that are always getting more and more precise. At present, experts believe that only 5% of the universe consists of visible material (stars, for example), and that it contains five times as much dark matter; the rest (i.e., 70%) is dark energy. In other words, we don't understand 95% of all that makes up the universe! In our initial model, we made sure to take this distribution of densities into account. We also used the map of inhomogeneities observable in the cosmic microwave background radiation (see the following box).

Figure 4.3 The importance of the Wiener filter.

The Anisotropy of Cosmic Microwave Background Radiation

"Cosmic microwave background radiation" refers to a radiation, first observed in 1964 by Arno Penzias and Robert Wilson, that suffuses the whole universe. According to the standard cosmological model, it was emitted 380,000 years after the Big Bang, when the first photons escaped from the hot (approximately 2,700°C), dense, and homogeneous "primordial soup." Since then, this "fossil" radiation has been cooling due to cosmic expansion; today, it reaches us at a temperature of 3K (that is, at −270°C). Its detection represented the first experimental proof for the Big Bang. Observation of a dipole in the homogeneous radiation provided evidence

−500 ━━━━━━━━━━━━ 500 μK_{CMB}

Figure 4.4 Cosmic microwave background anisotropy. The map on top shows the dipole due to our galaxy's motion; removing it allows us to detect the inhomogeneities due to primordial fluctuations (bottom map).

(continued)

----

that we are moving through the universe at a speed of 630 km/s, and pointed toward the existence of the Great Attractor.

At the beginning of the 1990s, the COBE (COsmic Background Explorer) satellite was launched to obtain more precise measurements in every direction. Indeed, once the dipole due to our own motion is subtracted, the cosmological model predicts that this radiation will not be perfectly homogeneous; instead, it should present tiny thermal fluctuations on the order of 1/100,000—signs of the overdensity necessary for the large structures we observe in the current universe to have germinated long ago. The mission proved a success: as expected, it showed that cosmic microwave background radiation differs depending on the direction in which one looks ("anisotropy," in professional jargon), thereby proving that these small clumps of inhomogeneous (dark) matter were present in the primordial soup. The mission's directors were awarded the Nobel Prize in Physics in 2006. Since then, other projects in space have sought to map tiny fluctuations in the cosmic microwave background radiation, and they have produced better and better resolution through increasingly precise temperature measurements. In the 2000s, we used data from NASA's WMAP satellite for the CF1 and CF2 maps; more recently, for the CF3 data set, we've been working with data provided by the Planck satellite operated by the European Space Agency.

This map represents a fundamental element of our guiding model, for the small fluctuations of energy density in the homogeneous primordial soup that constituted the universe when it was young are presumed to be the seeds of complexity responsible for forming the galaxy structures we observe today. More specifically, our initial model works with the spectrum of forces that produces the mean spatial separation between two inhomogeneities, and their relative intensities. It's important to bear in mind that the spatial distribution of masses is what gives each cosmic element its velocity.

Based on these factors—the standard cosmological model (as a working hypothesis) and data we have concerning a few peculiar radial velocities (for plotting purposes)—our software reconstructs vectors at fixed positions distributed at regular intervals for a given region, which it then aggregates and "smooths out" to obtain a velocity field. In turn, on the basis of this velocity field, the program derives a field of mass density. (After all, the gravitation exerted by matter is responsible for the peculiar velocities of galaxies.) A good indication of our model's success was provided by comparing this field of mass density, which basically consists of dark matter, with the positions of 100,000 galaxies known to exist in a part of the universe that had not been used in calculations. Comparing the dynamic and static maps, and seeing that they matched up, confirmed the reliability of our analysis.

All in all, five years passed from obtaining our CF1 sample of 1,800 dynamic galaxies until we published *Cosmography of the Local Universe*, a set of 22 maps made with the reconstruction program we'd developed. It had been worth pressing on until we were completely satisfied! In effect, this was the first, and nearly exhaustive, survey of our extragalactic surroundings. To our great relief, processing the CF2 dataset—which led to the identification of Laniakea's borders—went much faster (two years), because the methods employed had already been tested and proven to work.

Dark Matter

Astronomers basically have two ways to detect the presence of matter and determine its quantity. They do so either by measuring luminosity (luminous mass), since the quantity of light that a celestial body emits depends on the mass that produces it, or by measuring movement (dynamic mass): mass creates a gravitational field affecting the velocities of surrounding objects. Kepler exploited this principle when he calculated the dynamic mass of the sun simply by studying the revolution of planets in our solar system.

Later, questions arose when astronomers compared the quantities of mass estimated by these two methods. In the 1930s, Swiss astronomer Fritz Zwicky was the first to recognize that the dynamic mass of a galaxy cluster far exceeds its luminous mass—a surprising discovery that passed almost unnoticed at the time. In the 1970s, Vera Rubin also observed a great difference—by a factor of ten—between dynamic mass and luminous mass, this time within spiral galaxies themselves. In either case, luminous mass proved much weaker than dynamic mass, suggesting the presence of something that telescopes cannot detect; this invisible mass was given the name "dark matter."

The true nature of dark matter remains a mystery that many particle physicists are hard at work trying to solve. It's unlikely to be ordinary matter (sometimes known as "baryonic" matter), since it produces no visible light. Instead, it's more likely to be "exotic" in origin—and, moreover, not necessarily a substance of uniform nature (it differs depending on whether it's found in the Rubin Galaxies, Zwicky Clusters, or out in extragalactic space). Cosmographers need to consider fluctuations in extragalactic dark matter to explain the formation of large structures, which couldn't have come about through visible matter alone.

The theory goes that there are two types of extragalactic dark matter: hot dark matter and cold dark matter. Hot dark matter consists of particles that are light and move rapidly; this suggests

(continued)

(continued)

a "top-down" scenario for the formation of large structures: first, superclusters formed, then they broke up into clusters, which then brought forth galaxies. Cold dark matter is composed of heavier, slower particles, which implies a "bottom-up" scenario: galaxies formed first and then grouped together to form clusters, and finally superclusters. According to current research, the latter scheme seems more probable. Either way, however, the mysterious matter in question still hasn't been detected. Doing so is the purpose of projects such as EDELWEISS, an experiment being conducted in the Modane laboratory housed in the Fréjus tunnel between France and Italy; here, researchers are tracking WIMPs (weakly interacting massive particles)—prime candidates for cold dark matter.

The Worlds According to Daniel

Daniel Pomarède is a physicist working at the CEA Institute of Research into the Fundamental Laws of the Universe, Saclay (near Paris). With a PhD in cosmology and a specialization in antimatter, he has devoted nearly a decade to programming highly sophisticated visualization software: "Saclay Data vision," or "SDvision" for short. This software consists of some 100,000 lines of code, which he is constantly working on, in keeping with new developments in the fields of astrophysics and particle physics. The programs are always being improved. In December 2010, he met Brent Tully at a conference in Ouagadougou, the capital of Burkina Faso. There, Pomarède was demonstrating how his software is suited to representing diverse scenarios of galactic formation on static maps.

Vera Rubin

Figure 4.5

American astronomer Vera Rubin was born in 1928. When denied the opportunity to earn a master's degree at Princeton—which didn't admit women to this program until 1975—Rubin first enrolled at Cornell, then completed her doctoral studies at George-town. Her dissertation, which she defended in 1954, concluded that galaxies are mostly distributed in groups throughout the

(continued)

(continued)

universe, not at random; however, this idea didn't receive serious attention for the next twenty years. Rubin made her most famous discovery by measuring the orbital velocities of stars in our galaxy. In 1962, she published a study of the velocities of 1,000 stars. Stars at a distance over 28,000 light-years from the galactic center display a constant velocity, even though, according to Kepler's laws, their velocity should decrease as their distance from the center increases. Rubin concluded that an invisible mass is present in peripheral regions of the galaxy. Efforts to explain this phenomenon of our galaxy's rotation led to the theory of dark matter. Further observations—among others, of the bullet cluster, gravitational lenses, and galaxies' motions within clusters—have lent the hypothesis further support. In 1965, Vera Rubin became the first woman authorized to use the telescopes at the Palomar Observatory (United States). In 2002, she received the Gruber Cosmology Prize, the highest international distinction in the field in recognition of her pioneering work in understanding gravitation and all types of matter.

Tully had long recognized the importance of high-quality visualization techniques. After all, our field is a matter of designing maps, which should be as clear as possible! Moreover, at the end of the 1980s, he'd created a color atlas of the Local Universe— a limited-print edition that professionals regard highly. Tully acknowledges a host of reasons for presenting our discoveries in the best possible way. Above all, effective representation facilitates the transmission of new findings, both to the public and to researchers, and that's the main purpose of our work. I think he secretly dreams of the day when anyone on our planet will be able to locate the border between the Perseus-Pisces Supercluster and Laniakea as easily as they might go to the shop around the

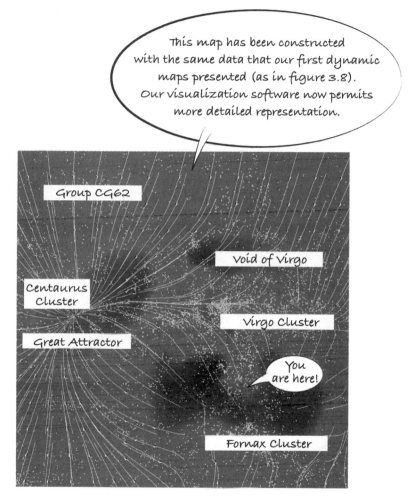

Figure 4.6 The purpose of visualization. (Also plate 8.)

corner … Brent also believes—and rightly so—that sophisticated schemes of visualization help us to analyze data and understand the causes and effects of physical laws governing the organization of large-scale structures. This involves being able to take a virtual walk through our maps, changing the scale or angle of views at will, simultaneously displaying different physical variables (fields of density or velocity, visible matter or dark matter), and simulating galactic flows in gravity wells …

When he saw Daniel's presentation, Brent immediately recognized the immense potential of software like SDvision. For his part, Daniel was drawn to the Cosmicflows project. After the conference, Brent told me about their encounter, and that's how we found a new member for our team. Ever since, he's been blowing us away with his increasingly creative ways of visualizing samples that are only getting bigger!

The Countdown Begins

Ultimately, by expanding our team, we were able to broaden our cosmic horizon, too. We formed a group in which each member's expertise ensured the success of every step taken. Now, we'd be able to locate the position of any number of galaxies in the Local Universe; for some of them, we could also determine peculiar radial velocity, which made it possible to trace them; on this basis, we'd be able to reconstruct an underlying density field of dark matter and the trajectory of galaxies, thereby defining the limits of grand-scale structures.

We had all the tools we needed for redefining the contours of our supercluster. Spring 2013 went by in a flash. On 28 April, Brent shared a homogenized version of the Cosmicflows-2 data set. The new information was plugged into Yehuda's software.

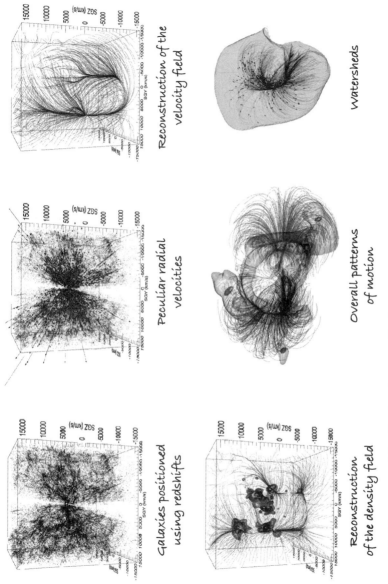

Figure 4.7 Maps summarizing the steps in our method.

On 14 May, I sent the results of analysis to Daniel for visualization processing. Two days later, Brent arrived in Paris to admire the results; both of them watched as galactic currents took shape in a sea of dark matter. On 20 May, Brent arrived in Lyon, maps in hand; he'd even managed to avoid being sidetracked by the charms of wine country. The dynamic maps we'd devised presented incontrovertible evidence: before our eyes, clear lines of separation and sinuous galaxies surrounded our supercluster. Brent and I took advantage of the week he spent in Lyon to write the article we'd finally submit, almost a year later, to *Nature*.

In the course of that year, we worked to refine our maps on the basis of more complete versions of the CF2 data set, using different parameters to reconstruct the velocity field and visualizations with varying points of emphasis and perspective. Above all, we spent a great deal of time discussing the merits of our new way of defining a supercluster—first within the group, and then with colleagues at conferences. Choosing a name for our discovery took a while. At first, we thought of calling it "Aina Aina," which means something like "the house in the stars" in Hawaiian. In the end, we chose "Laniakea"—"immense heaven"—a more appropriate title, which has the further advantage of being easy to pronounce in all languages.

For several decades already, researchers had been able to determine the positions of galaxies and represent them cartographically—thereby demonstrating how they form clusters grouped along filaments, with "empty" zones in between. Such static maps had revealed a whole network of large-scale structures, but the borders remained vague. In particular, structures that were dense with galaxies had come to be called "superclusters," but the term wasn't clearly defined.

In spring 2013, we cleared a new path for analyzing large-scale structures. Doing so represented the third, decisive step (after opting to focus on spiral galaxies and using the Wiener filter) that led to Laniakea's discovery. On the velocity field we'd calculated with the Wiener filter, we mapped out currents of matter. Then we looked for areas where they met up and, even more importantly, areas where they separated. These zones of divergence amounted to "dividing lines": on either side, the cosmic flow of galaxies took opposite directions. By connecting all the points where divergence occurred, we obtained a surface. All galaxies located within the volume this surface delineates move toward its interior. In this context, we proposed a new definition of the term "supercluster": the volume contained within such a surface. The first supercluster for which we determined the entire volume is the only one that lies entirely within the area mapped out with our Cosmicflows-2 sample. This is the supercluster to which our own world belongs: Laniakea.

Only after subjecting our results to much further analysis, which strengthened our belief that we had really discovered something big, did we submit the article to *Nature*—on 18 April 2014. Three months passed before it was accepted, then another two before it was published. Research projects always culminate in an article detailing the experimental protocol, data used, methods of analysis, and conclusions. Once written, the article is submitted to the editor of a professional journal, who asks one or more anonymous readers—referees chosen among the top specialists in the field—to assess the project and findings. Correspondence between readers and authors follows; the process is often lengthy and sometimes fraught, but always fruitful. Such review is absolutely necessary to confirm the reliability and relevancy of claims. One can never be sure of the outcome: following months of discussion, the article may wind up being accepted or declined. This system of exacting and anonymous judgment by peers ensures that researchers produce rigorous work. After all, scientists aren't magicians. They have to explain their tricks! One side benefit is that such discussions often lead to new ideas for future projects.

Figure 4.8 shows a representation of Laniakea made on the basis of the Cosmicflows-2 data set. Small white dots indicate galaxies. The underlying density fields of matter display a color according to relative strength (red-orange for high density, blue to dark blue for void areas). The flow lines of galaxies are represented in solid white when they belong to Laniakea; otherwise, they are gray. Laniakea's "dividing line" appears quite clearly, in

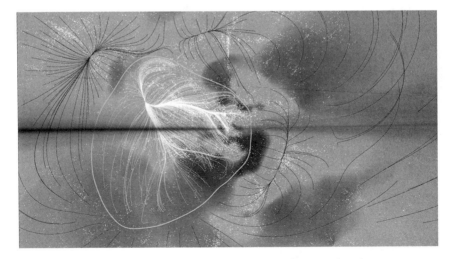

Figure 4.8 Evidence of our Laniakea Supercluster through a visualization from the CF2 dataset. (Also plate 9.)

orange, and distinctly shows where the supercluster's distribution surface intersects with the plane of the map. We can see a
dark band stretching out horizontally: this is the area obscured
by our own galaxy (the Zone of Avoidance). Our software has made
it possible to connect galactic currents on either side of this hidden region.

The Watersheds

Laniakea measures 500 million light-years across, but its "center"
is around 250 million light-years from us. Unsurprisingly, the
Centaurus Cluster is located where five large filaments meet; as
such, it represents the point of convergence for five distinct cosmic flows. It would seem that the position of the Great Attractor
has finally been found, then. The site, in plain view, looks more
like a small, flat valley—into which galaxies are pouring because
of gravitation—than some gigantic, invisible mass. Well, there it
is! Now we have a clear picture of the supercluster where we live.

A simple analogy will help to make this new way of representing superclusters comprehensible in concrete terms. Imagine that
galaxies are drops of rainwater following their normal course. The
line dividing two superclusters—that is, the boundary where galaxies head off in one direction or the other, in keeping with gravitational pull—corresponds to the watershed between two drainage
systems, or the ridge line on a mountain. On Earth, wherever
ridge lines form a closed loop, they demarcate a watershed—
which, in the extragalactic perspective adopted here, corresponds
to a supercluster of galaxies. After all, "watershed" names an
area where all flowing water—from tiny streams to large rivers—
converges toward the same outlet (often a sea). In the same way,
all the galaxies in a supercluster travel down the same planes or

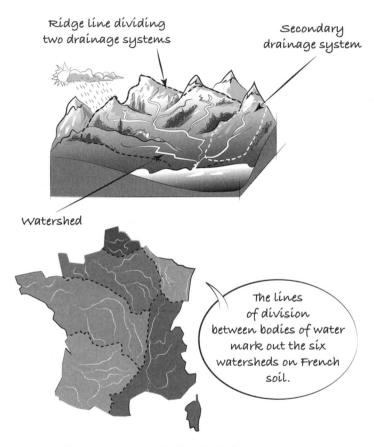

Figure 4.9 Illustration of watersheds in hydrology.

filaments, drawn along by a single, overdense center of matter. Like superclusters, watersheds can occupy vast expanses: the territory of France can be divided into six principal drainage systems, from which emerge the main rivers of the land.

It stands to reason that the first extragalactic watershed we identified and defined was the one containing our own galaxy. The structures are immense, and the borders between two superclusters often occur at the edges of our maps. Most of the superclusters surrounding Laniakea are partially visible in figure 4.8, including Perseus-Pisces and Coma (the Coma Berenices Cluster). Pavo-Indus is located behind this projection and therefore not visible. The relatively unexplored Shapley region can also be seen, which promises to reveal a gigantic supercluster at some point.

Our galaxy is located at the edge of Laniakea, close to the border with the Perseus-Pisces Supercluster. This means that some galaxies, even though they are close to us, belong to a neighboring supercluster; attracted by another point of convergence, they are drifting in a direction opposite to our own. Our proximity to Perseus-Pisces allows us to define, at least approximately, the surface englobing this supercluster. That said, it should be noted that its rear limits are still poorly defined, since we lack observations in this area. We'll just have to continue to observe further galaxies.

Very Peculiar Motion

I'd like to finish this guided tour of our cosmic surroundings with a reminder that the distances at issue are phenomenal; the universe's expansion, which tends to draw galaxies apart—and proves stronger the greater the distances between them are—is more powerful than gravitation, which, for its part, generally draws objects together. Our maps, however, privilege gravitation; we have deliberately bracketed expansion in order to identify objects' peculiar motion.

It's important to remember that our diagrams don't display galaxies' motion as a whole. Indeed, the large-scale flows of galaxies we have described represent mere disturbances when compared to the universe's expansion: all galaxies, with the exception of ones that are extremely close to each other, neighboring clusters, and groups, are moving apart. We made it a point not to define a supercluster as a "linked gravitational structure," since its components are actually in the process of drawing farther and farther apart! On this score, we can see the limits of making an analogy between hydrology and astronomy in order to visualize the extragalactic environment. To take cosmic expansion into the picture, we need to imagine the earth's crust swelling rapidly—like a big balloon being inflated. In this picture, water from a small mountain stream still is flowing into the valley, but it never reaches the bottom; the bed of the stream is expanding much too fast...

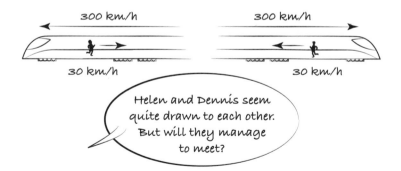

Figure 4.10   The train analogy to help clarify Laniakea's future.

A Supercluster Is Born

As noted previously, Laniakea means "immense heaven" or "immense heavenly horizon" in Hawaiian. It represents a *horizon*, since discovering the watershed divide that surrounds it made us aware of its existence in the first place. And it's *heavenly*, of course, because it's located in the heavens. Finally, *immense* refers to the fact that it's one hundred times the size of what was previously defined as a supercluster. We chose the name to honor the Polynesian navigators who steered their way using the stars. Nawa'a Napoleon, a professor at the University of Hawaii,

Figure 4.11 A new line in our cosmic address.

first suggested this title, which was then officially approved by the members of the International Astronomical Union.

Right away, our discovery met with widespread enthusiasm. The article on the *Nature* website was read over 20,000 times by researchers. In just a month, the video accompanying it became the most popular science clip on *Nature*'s YouTube channel. It's still one of our most viewed videos, with more than 6 million hits to date. Laniakea has been printed on T-shirts and scarves. We can listen to more than 100 musical items named after Laniakea. There's even a video game with a superhero called "Laniakea, the eternal navigator." Laniakea medallions have

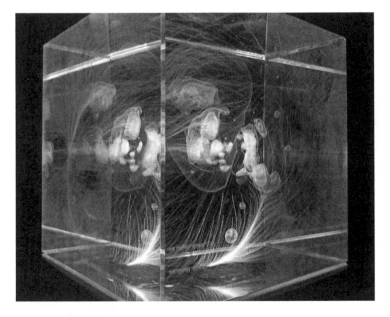

Figure 4.12   Laniakea enclosed in a glass cube.

been manufactured; recently, an artist even sent me three-dimensional models of our maps made out of glass! It's a shame that none of the profits from these products has gone to finance our research. All the same, such sharing—when a broader public takes up scientific findings—adds a further dimension of meaning to what we do. This is how the knowledge of humankind as a whole grows every day.

For our own part, we scientists participate in events that are free of charge and open to the public. For instance, I was involved

Figure 4.13 An image of Laniakea projected onto a building façade, Lyon Festival of Light, 2014. (Also plate 10.)

in the Lyon Festival of Light in December 2014, which millions of visitors saw as they walked down the city streets. Other activities have included Researchers' Night, the Festival of Science, and events at schools. Our maps made the covers of several magazines—including *La Recherche*, which listed Laniakea among the "10 discoveries that changed the world in 2014" for all the sciences, internationally.

5 Beyond Laniakea

The continuation of the Cosmicflows program: improving our maps to advance the understanding of the universe's physical phenomena.

Virtual Universes

While conducting observations and analyzing data to create our large-scale maps of galactic motion, I pursued my interest in a seemingly more abstract field: the numerical simulation of constrained systems. In 2011, when I returned to Europe after another year working in Hawaii, I set out on entirely uncharted terrain—for me, at least! Some friends prone to worry probably feared that I'd get lost flitting from one method to another. But my own thinking was the opposite: because I was exploring something unknown, I should use all available means to get a truly comprehensive perspective. In my quest for the Great Attractor, I've become convinced that no stone should be left unturned and all leads are worth following—even if, in the end, they don't go anywhere. Plus, in purely intellectual terms, it's quite gratifying to acquire new skills in other fields of research.

What, then, is "numerical simulation"? As the name indicates, it's a kind of illusion—or, more accurately, an imitation of reality. It's a matter of using theoretical models to reproduce experimental observations as faithfully as possible. Many fields of research employ numerical simulations; in my case, the point is to devise maps that mimic the universe. But what's my real motivation for making these counterfeits? Even if we're proud of our maps, they aren't unique or fragile objects to be guarded like works of art. And although some mercenary entrepreneurs have already laid hold of Laniakea to sell any number of products, the whole business of getting rich off fakes doesn't interest me. The purpose of my simulations isn't duplication. Instead, simulated maps based on theoretical hypotheses are made to be compared with "real" maps based on empirical observation. When the two kinds of map match up, it confirms a whole series of choices we've made in our research program.

Numerical simulations require a computer, the ideal tool for performing complicated calculations quickly, without error, and repeatedly. In our case, the calculations are so numerous and complex that we need supercomputers. In fact, these extremely powerful computers are quite rare (each country has one or two on average), and the number of scientists who want to use them is constantly growing. As a result, researchers often find that they have to stand in line—just like observers of the sky, who have to wait for telescope time. Our team managed to gain access to two supercomputers, the SuperMUC in Munich and the Mare-Nostrum in Spain (among others). But since these computers are never reserved for just one project, it usually took us one or two months of calculations just to complete one simulation!

After all, simulations of the universe can't be left to chance. It doesn't take a computer that's "super" simply to place 100,000 small dots (representing galaxies) at random in a volume 350 million light-years in radius (our Local Universe). Our numerical simulations involve an immense amount of detail: we provide numerous parameters, which represent key ingredients and steps for making our cosmological recipe a success. When calculating our virtual maps, we base our work on the same hypotheses used to reconstruct velocity fields with the Wiener filter. These parameters are grouped together in what's known as the Standard Cosmological Model. We employ the latest data provided by the Planck satellite: the most likely distribution of energy (dark energy [68%], dark matter [27%], and visible matter [5%]), as well as the spectrum of small density fluctuations in cosmic microwave background radiation.

In the small world of cosmologists, there are many international teams conducting numerical simulations like this. Over the last two decades, research has confirmed observers' impression of the "cosmic web": on a grand scale, the universe is laid out as if a spider had spun it, and galaxies are found in the planes, filaments, or nodes tying together large, empty spaces. It's important to bear in mind that this scheme of organization—which, although recent, is almost common knowledge already—isn't easy to observe in actual fact. Researchers have been mapping, as precisely as possible and with great difficulty, just the tiniest fraction of the observable universe—one millionth of its volume. Our view of the universe on a grand scale is provided only by numerical simulations, which improvements in computing have made possible. At this level, we only have artificial representations of the cosmic web, calculated on the basis of the Standard Cosmological Model.

As when making "real" maps based on observations, the best methods should be employed when designing "fake," simulated

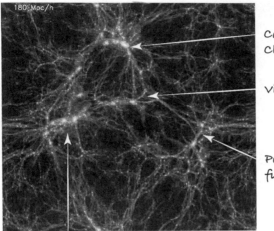

180 Mpc/h

Coma Berenices
Cluster

Virgo Cluster

Perseus-Pisces
filament

Great Attractor

Figure 5.1 The cosmic web obtained through a CLUES numerical simulation.

maps. Those that were available at the time proved too "unreal": they depicted a universe I didn't recognize. For this reason, I got in touch with a group of German and Spanish theoretical computer scientists led by Stefan Gottlöber and Gustavo Yepes. Together, we developed a new way of conducting numerical simulations, which, this time, would be "constrained" by our own, empirical findings. The task was to improve the cosmological recipe for cooking up a universe by adding observational parameters to the process of calculating the position of virtual galaxies. These additional constraints themselves had nothing virtual about them; they represented the actual positions and velocities of galaxies used in our maps. The approach is very different from the conventional method. We're the first researchers in the world to have developed it.

In modeling the events that followed the Big Bang, we weren't content simply to make our point of departure the almost-uniform primordial soup, with its assorted clumps of inhomogeneity. Instead, our calculations for constrained simulations incorporated a hint of dark matter. In concrete terms, this meant working with Timur Doumler and Jenny Sorce (doctoral students at the University of Lyon) to code a sort of time machine. The algorithm that we devised took into account the positions and velocities of galaxies in the Local Universe today, actual values we had mapped out. On this basis, the program calculated where they might have come from billions of years ago, as well as the velocity they had then. We assigned this "velocity of olden times" to particles of dark matter at the outset of the simulation, because it likely represents the speed at which fledgling galaxies moved during the universe's infancy. Once this step had been completed, we launched a standard numerical simulation, then sat back to watch the history of the universe unfold in the correct chronological order. Before our eyes, galaxy structures were born and grew up.

In contrast to the "classic" form of numerical simulation—which generates the velocities and positions of initial fluctuations at random and therefore yields a simulated universe that differs from our extragalactic environment—our constrained simulations aimed to produce a map of the artificial universe that admits direct comparison with observed maps. The results proved very convincing: we'd simulated a very young universe, in keeping with the data supplied by the Planck satellite, which evolved into a version of our Local Universe as we understand it on the basis of large-scale structures that we have observed. The accordance was immensely gratifying, for it confirmed our hypotheses. Indeed, changing the parameters of the simulation— for instance, entering a different age for the universe or another

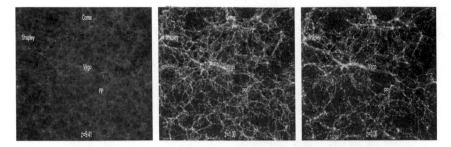

Figure 5.2 Time-lapse produced by a constrained simulation based on CF2. Here we can see the birth and development of our corner of the universe, from left to right at the ages of 500 million years, 6 billion, and 13.8 billion years. (Also plate 11.)

proportion of dark matter—yielded large-scale structures that weren't the same as the ones on actual maps.

In fact, each simulation comprised a series of maps calculated for an array of dates in succession, from the time of the universe's extreme youth (as evident in cosmic microwave background radiation) up to the present. By passing this collection of maps in quick review on a video monitor, we were able to visualize the whole history of the universe in just a few seconds. Large structures, such as Laniakea, seemed to take distinct form pretty fast—after some four or five billion years. Especially on a small scale, our simulations revealed, in astonishing fashion, the formation of the Virgo Supercluster, the birth of the Great Attractor region, and galaxies streaming away from areas that later became voids.

Given the current state of knowledge, the Big Bang theory offers the most satisfactory model for understanding astronomical observations. It received this title in the 1950s from one of its main detractors, the British astronomer Fred Hoyle, who advocated the Steady State model instead. Erroneously, the term "Big Bang" is often taken to refer to the moment when a massive dilation of space occurred, 13.8 billion years ago. One shouldn't confuse this extremely rapid inflation of the cosmos with the expansion, at a much lower rate, that Hubble and his colleagues identified (which the book at hand has been discussing at such length).

Before it inflated, the universe was extremely dense and hot—so much so that all its elements were combined, as it were: time, energy, space…We don't know how to describe the state of the universe before this point; the laws of physics as we know them don't apply. After inflating, the universe cooled down so much that the four fundamental forces separated; then protons and neutrons, the future components of atoms, formed. A second after the Big Bang, matter had already won the battle against antimatter, a fight between warring siblings different only in terms of their electrical charges; this combat proved unequal due to the (very small) difference in numbers. After three minutes, the first, airy nuclei formed: deuterium and helium. This primordial synthesis determined the current mass proportions of these atoms: 73% hydrogen and 25% helium.

The universe continued to expand—that is, to cool—until it reached the age of transparency. After 380,000 years, light finally managed to escape from the primordial soup, and electrons paired with nuclei to form the first atoms. It is after this point that astronomers are able to look into the past by staring off into the sky, toward the outer limit of what's known as the "observable universe." As we remarked earlier, it's impossible to look back beyond this point, because the universe was opaque; the depth cannot be sounded with telescopes. Satellites such as Planck give us a glimpse, in the nearly homogeneous soup, of little clumps of dark matter that formed around primordial quantum fluctuations and, in turn, collected new, "ordinary" matter. Dark matter was always there, playing a hidden role in the formation of galaxies. After several billion years (according to our simulations), the galaxies grouped into clusters, and then superclusters—like Laniakea.

A Berlin Ballad

The name of our project of numerical simulations is CLUES: Constrained Local UniversE Simulations. As in any collaborative undertaking, ideas are constantly being exchanged. Needless to say, videoconferences serve this purpose well. All the same, we still need to meet in the flesh to discuss things more concretely. Thus, once a year, we organize a weeklong session at one of the universities where team members work. Locations include Spain, France, Germany, Israel...

One of the driving forces in CLUES is Stefan Gottlöber. For many years now, he has been working at the Leibniz Institute for Astrophysics at Potsdam—a city with 150,000 inhabitants just outside of Berlin. This charming place, with a large river running through it and dotted with lakes and forests, is home to many palaces built when Prussian royalty had their seat here. It's also known for having been the site of the Potsdam Conference in 1945, when the Allies decided the fate of the countries that had lost the war. Subsequently, Potsdam wound up belonging to East Germany. The situation really became difficult when the Berlin Wall went up and travel between Potsdam and West Berlin was prohibited. The city's residents could only reach East Berlin by taking a detour of more than 70 kilometers.

I'll never forget my first visit. Stefan was my guide on the country roads. He'd stop regularly—practically every kilometer—get out of the car, and, walking from one side of the road to the other, gesture wildly to draw a border that no longer existed. He seemed to have difficulty believing that it had really been there. "This was Potsdam, and that was Berlin—the East over here, and the West over there! And here's the famous Glienicke Bridge, where Russians and Americans traded spies..." As I sat in the car, watching him striding back and forth in a frenzy, I realized just how lucky

I'd been: to have grown up in a land that guarantees liberty (and, I hope, will continue to do for some time yet). By transforming a simple country lane into an iron curtain, Stefan brought to life the absurdity of a world cut in two. I appreciated his disbelief at living in Potsdam today—after waiting for forty years to have the right to cross the road. When I asked what had changed the most for him after reunification, he didn't have to think at all: freedom of movement. Previously, he'd only left the GDR a few times, to visit Czechoslovakia. Today, with incredible gusto, he takes advantage of his freedom and the opportunities our profession offers to see the world. I've even been his guide on trips to Lyon and Honolulu, so I'm quite sure: Stefan welcomes every chance to experience something new!

I Want More!

But let's get back to physics. As we've seen, we're living in a galaxy supercluster that's one hundred times bigger than anyone thought fifty years ago. But in science, any answer prompts new questions. For example, is the size of Laniakea—some 500 million light-years—something typical? Is our supercluster big, or small? To answer these questions, we need to be able to measure the dimensions of at least one hundred other superclusters. On that score, our constrained simulation, which is the same size as our map based on observation, can't tell us anything...

We've also seen that our galaxy is located close to the dividing line between Laniakea and the neighboring supercluster of Perseus-Pisces. The latter seems to have similar dimensions; perhaps it's a little larger. The volume we have mapped includes its "center," but we're unable to visualize the lines of separation at the greatest distance from us. The velocities of the most remote galaxies we've measured, just beyond the center of the

Perseus-Pisces Supercluster, all point back to it. We still can't see far enough to make out whatever galaxies are moving away from Perseus-Pisces, toward the next watershed. How annoying! Our galaxy sample needs to go deeper.

In the opposite direction from Perseus-Pisces—at the other end of the map, in the line of sight of the cosmic microwave background dipole and far beyond the Great Attractor—we've spotted a concentrated mass of thirteen large galaxy clusters. Here, too, we need more data to confirm that the galaxies located farther away are, in fact, flowing toward these clusters, which evidently form the central valley of a gigantic supercluster, Shapley (the only large structure named after its discoverer). We can tell that we're getting close; we have the right methods in hand and the right tools, too. But we still need to expand the volume of our maps on all sides. That was the goal of the next part of our research program: compiling data that would yield Cosmicflows-3. Doing so promised, among other things, to allow a large part of what appears to be the Shapley Supercluster to be included.

The new Cosmicflows-3 program was built in four years. In order to conduct observations at a greater distance, we used data from the giant radio telescopes in Green Bank and Arecibo for extremely long exposures and obtained over 5,000 new spiral galaxy measurements.

We also used the NASA Spitzer space telescope (which can be seen in the gallery in figure 3.3). A telescope in orbit, outside the earth's atmosphere, provides images that are much clearer, which means that we can observe objects that are farther away. Another reason for choosing the Spitzer satellite was the fact that it enabled us to conduct observations in the near infrared range. In other words, we could see through the dust clouds of our own galaxy and get much better results than when using an ordinary

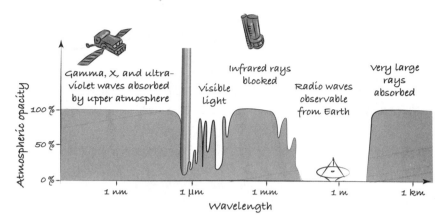

Figure 5.3 Absorption of electromagnetic waves by the earth's atmosphere.

optical telescope. Observing the extragalactic universe closer to the disc of our own galaxy at this level permits us to reduce the impact of the Zone of Avoidance on our maps. That said, the closer we get to it, the more the stars in the foreground, which belong to our own galaxy, intrude on the images. So there's always something that we can't map, even with this satellite... To date, our observation campaigns with Spitzer have enabled us to collect images and photometry in the near infrared range for over 4,000 spiral galaxies. Some of them had already been measured; but additional measurements are never useless: often, the quality of parameters is more important than quantity.

Finally, Cosmicflows-3 also includes a large amount of data obtained by another team. The researchers who provided it, based in Australia, made a series of distance measurements for 8,000 elliptical galaxies in the southern sky; they did so using the Faber-Jackson method (dating to the glorious era of the Seven Samurai) and a multifiber spectroscope installed on the good old Schmidt telescope at my beloved Coonabarabran Observatory. Their device is called the 6dF (6-degree Field), which refers to the field of vision it measures. The technique employed is similar to the one I developed when I was working on my thesis there. Twenty years later, everything came full circle! But we'll need to observe the greatest caution in evaluating this sample of elliptical galaxies. When using measurements made by other astronomers—especially ones obtained by means of the imprecise Faber-Jackson method—it's always better to make sure that they're sound.

In 2016, the Cosmicflows-3 data set was made available to researchers worldwide (and to the public, too, since all our findings can be consulted online). It was a wonderful way to celebrate the tenth anniversary of the project. CF3 contains measurements of distance and velocity for about 18,000 galaxies, covering a radius of 600 million light-years (Mly) from our position. As a reminder: the Cosmicflows-1 data set, which was published in 2008, included 1,800 galaxies within a 130 Mly radius; in turn, Cosmicflows-2, which appeared in 2012, contained 8,000 galaxies within a radius of 350 Mly. Our knowledge of the Local Universe has come a long way in a decade. It gives me a great sense of satisfaction to see such progress!

Toward a Homogeneous Universe?

Even though CF3 data is already available, it's still far from having been interpreted in full. In the volume of universe explored by Cosmicflows-3, we expect to see the contours of a dozen superclusters at most. Preliminary analyses suggest that we won't be able to map the entirety of the remote and imposing Shapley watershed. But all the same, the CF3 data set provides a great deal of local information that will allow us to shore up our overall understanding of the universe, largely by analyzing the motion of nearby structures.

Every day, every morning, we arrive at our desks eager to make some new discovery. Analyzing the velocity maps obtained through Cosmicflows-3 has enabled us to offer a plausible explanation for an intriguing structure. On the map of cosmic microwave background radiation that the Planck satellite has provided, we can see a region that is cooler than elsewhere. Cosmologists call this area the "Cold Spot." If modern cosmology—that is, the sum total of physical processes (gravitation, the expansion of space, transparent matter, the age of the universe)—is correct, then the irregularities (clumps) on the map of cosmic microwave background radiation are "footprints" left behind by the large structures of the universe. Indeed, photons (light) emitted just after the Big Bang cut across galaxy filaments, voids, and superclusters to reach telescopes like the Planck satellite. The warm regions of the map represent marks left by massive structures, then, and cold regions the traces resulting from a lower quantity of matter.

Recently, we proposed that an especially empty region on our Cosmicflows-3 maps corresponds to the coldest region on the map of cosmic microwave background radiation for the whole universe. This has caused quite a stir in the scientific community. Like all advances, the result must be confirmed by other teams and independent experiments before victory can be declared. But if we're right, it would represent a major event for modern cosmology, since, as it stands, it's very difficult to connect, by means of observational evidence, the map of the universe's first few moments with the mature universe we inhabit.

When compiling Cosmicflows-2, we first measured the motion of our galaxy and its neighbors toward the Virgo Supercluster; then, along with this cluster, we measured motion toward the central valley of our watershed, Laniakea. For all the movements at issue, the overall velocity is about 560 kilometers per second.

Significantly, the dipole observed in cosmic microwave background radiation—a discovery made almost by chance over fifty years ago (which inspired all of my own research)—indicates that we are in fact moving at 630 kilometers per second, in a direction close to the center of Laniakea. The motion of galaxies (apart from the overall expansion of space) is due solely to the effects of gravitation on their environment. Picture a galaxy surrounded by others of the same mass, distributed evenly: it wouldn't move at all, since the same "force" would be pulling it in all directions at once. Now picture a tug of war, with contestants of equal strength on either side. The knot in the middle of the rope moves away from the side where the team slackens its grip. That's the image of the discovery we've made. The Great Attractor alone isn't enough to account for the velocity and direction of our galaxy's movement. For over forty years, astronomers had tried to understand the rate of movement at 630 kilometers per second, and we finally did something new.

Instead of focusing on which masses are pulling us in the direction we're heading, we decided—based on an excellent idea of Yehuda Hoffman—to look the opposite way: in the direction our galaxy is coming from. And what did we find? We recognized that there's a vast region without any galaxies, without enough "players" to make a team to pull the rope. To return to the image of the knot in the middle: its movement can be viewed in two ways; either it's being pulled by the bigger team, or it's being pushed by the smaller team. In keeping with this idea, we named the empty region the "Dipole Repeller." Remember that our galaxy's movement was first measured on the basis of observing a dipole on the map of cosmic microwave background radiation. We've made a video on the subject (see the filmography at the end of the book for a link).

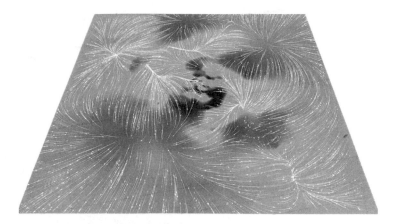

Figure 5.4 First map from the Cosmicflows-3 generation. Here we can see that the cosmic flows show the other watersheds surrounding Laniakea. (Also plate 12.)

With this new paradigm—attractors and repellers—a new vision of the universe's dynamic comes into view. Consistent flows are carrying our universe along, moving away from voids and heading toward large concentrations of mass. This gigantic architecture, whose parts we can identify by studying rivers of matter, constitute the "cosmic velocity web." The first map we've made of the cosmic web is publicly available. At the following address, you can wander the cosmic channels at your leisure with our interactive program: https://skfb.ly/667Jr.

At the same time, however, it seems that the notion of our world "falling" toward its watershed's interior fails to explain all of its motion. There's a small element of residual movement, on the order of 50 to 100 kilometers per second, that remains difficult to quantify, since the margin of error for velocity values on this scale lies at several dozens of kilometers per second. Accordingly, we need to observe extreme caution when analyzing data. But all the same, such residual motion suggests that Laniakea as a whole may also be moving—as if another, even larger supercluster were pulling at it. This view does not contradict our conception of the universe being divided into watersheds: it would have been too much to ask to find out that we're located right next to the center of the largest structure in the neighborhood… This hypothesis of "tidal movement" within an even larger watershed finds confirmation in the direction of the residual flow pointing toward the famous Shapley Supercluster. We can't wait to analyze the maps of Cosmicflows-3 more fully. If we're right, we'll be able to describe residual motion as a whole as the effect of attraction exercised by the Shapley Supercluster. Indeed, should the universe's volume—including Shapley—be immobile, we'll be able to understand, in more concrete terms, the size at which the universe can count as homogeneous and isotropic.

This is because the universe's homogeneity and isotropy represent the two pillars of the principle on which the cosmological model stands. According to this principle, no observer, wherever he or she may be in the universe, occupies a privileged position: no matter where one stands, one is surrounded by the same environment (homogeneity) and observes the same type of universe, irrespective of the direction of observation (isotropy). This principle was introduced in 1917 by Einstein, who needed the world to be symmetrical in order to solve the spatial equations of his theory of global relativity mathematically.

According to the cosmological principle, mass is uniformly distributed on a large scale; at the same time, this immense volume is drawn by gravitation in all directions at once. It follows that it should be immobile (apart from the phenomenon of

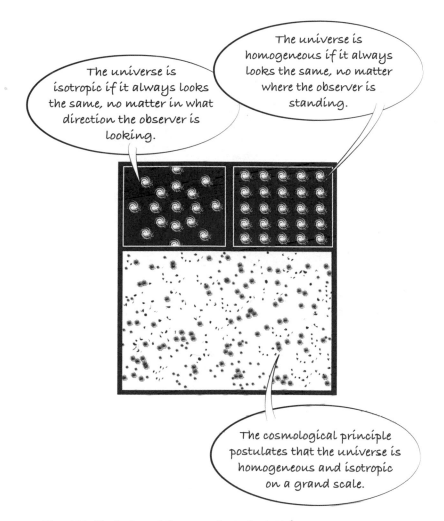

Figure 5.5 Illustration of the cosmological principle.

cosmic expansion). Experiments have verified the cosmological principle on an extremely large scale; in particular, it's evident when one considers the almost perfect isotropy of cosmic microwave background radiation. So why has so much space in this book been devoted to inhomogeneity and anisotropy? The reason is that, on a small scale—locally—the cosmological principle no longer holds true. We can observe any number of fluctuations in homogeneity, from "clumps of matter" in the soup 380,000 years after the Big Bang, to the large galaxy structures in today's local universe, and on to the cosmic microwave background dipole. Thus, thanks to our Cosmicflows-3 analysis, we feel we're close to answering a fundamental question: On what scale does the cosmological principle start to apply?

A Story without End

Knowing whether superclusters vary in size and understanding how they are organized in relation to each other promises to allow us to determine the critical scale at which the universe proves homogeneous. Such answers would let us understand even more about how large galaxy structures form and evolve. These lines of questioning are the very essence of the cosmologist's mission. But before we can answer, we'll need to determine the size of a significant number of superclusters, so we have a sample that's large enough to be statistically representative of the universe as a whole. As we've seen, the volume mapped by CF3 probably contains no more than ten of these extragalactic watersheds. The sample size is still much too small to be representative... That's why we're already planning Cosmicflows-4 and thinking about the features new telescopes should possess to satisfy our need to continue our journey beyond...

Across the globe, a new generation of telescopes is being built. In technological terms, it's proving more and more difficult to increase the diameter of giant radio telescopes without the structures collapsing under their own weight. All the same, a giant radio telescope 500 meters in diameter is in the process of sprouting from the earth in southwestern China: FAST (Five-hundred-meter Aperture Spherical Telescope), in the province of Guizhou. Exciting times are coming, since the ratio of useful diameters between FAST and Arecibo is a factor of 2.25. This will permit us to make significant advances, because this new radio telescope will also cover twice as much sky. Combining these two assets, we can expect to find 10 times more galaxies than before.

I'm putting more stock in completely new technology. Instead of increasing the diameter of a single telescope, the key is to synchronize and add together the signals collected by means of dozens of smaller dishes, each 12 to 15 meters in diameter. An ambitious project, the SKA (Square Kilometer Array), will start to be operational in 2018 and will be operational at 10% capacity from 2020 on. The idea is to bring together several thousand dishes in order to achieve an overall surface area for data collection that's about one square kilometer in size—which means that the SKA will be an instrument 50 times more sensitive than current telescopes. These antennas will also be spaced out, which means that the telescope will achieve excellent angular resolution: the images will be even clearer than those the Hubble telescope provides. Although an array of small dishes distributed over a small stretch (5 square kilometers) of the South African desert forms the telescope's heart, there will be others set up thousands of kilometers away, in western Australia. With these features and a vast field of vision, the SKA promises incredible results: we'll be able to study the sky up to 10,000 times faster than today! The SKA lies in the desert areas of the southern hemisphere in order to study the Milky Way (among other things), while avoiding interference from radio sources on Earth, which abound in the northern hemisphere because of industrialization there.

Programs to test the project's feasibility are already up and running. For instance, the APERTIF arrangement, located in the

Renée Kraan-Korteweg

Figure 5.6

A native of Amsterdam born in 1954, Renée Kraan-Korteweg has dedicated her career to the Great Attractor, focusing entirely on the region obscured by our Milky Way. Like many other astronomers, she travels extensively. Kraan-Korteweg studied in Basel, Switzerland. After receiving her doctorate in 1985, she worked in Groningen (Netherlands) and at the Paris-Meudon Observatory, before spending more than seven years in Mexico. During this period, she made many trips to conduct observations in other

(continued)

----

countries, especially Australia and Chile. Since 2005, she has lived in Cape Town, South Africa, where she directs the university's astronomy department. Kraan-Korteweg speaks six languages fluently. To observe the regions of the sky closest to the Zone of Avoidance, and even through the Galactic Plane, she has observed the sky across the whole spectrum: in the visible range, the near infrared range, the far infrared range, the radio centimetric range, and even in the domain of X-rays. Thus, in 1996, she found proof for the massive Norma Cluster (as large as 5,000 Milky Ways) and advanced the hypothesis that the Great Attractor will turn out to be a wall of galaxies cutting across the Galactic Plane; the huge cluster she located, officially labeled Abel 3627, forms its center. Ever since, Kraan-Korteweg has been working tirelessly to confirm the size of the "Norma Wall." She's a model of determination. Even more recently, she and her team published a data set of 883 galaxies hidden in the Galactic Plane, based on measurements carried out at Parkes, Australia—one more vital contribution to our understanding of motion in the Local Universe.

Netherlands, is testing technology known as a focal-plane array, which expands the field of vision by a factor of 25 by multiplying the number of collectors on the dish's focal plane. The SKA's precursor, MeerKAT, already has 64 antennas at the South African site, and the ASKAP telescope in Australia, equipped with 36 antennas 12 meters in diameter, is operational, too. These multitelescopes have started to receive the first extragalactic signals; soon, we'll begin the observation campaigns that will enable us to compile the Cosmicflows-4 data set. In particular, we're working with the Wallaby program on the ASKAP telescope, which is focused on measuring 21-centimeter hydrogen lines emitted by spiral galaxies; ultimately, the study will cover two-thirds of the sky and measure redshifts for 600,000 of these galaxies! Our own project aims to have mapped the distance and velocity of more than 50,000 galaxies situated in a radius of over a billion light-years in the course of five years (by my estimate); this corresponds to a volume possibly containing several hundred superclusters like Laniakea. The prospect of being able to answer a host of riddles in the near future is quite exciting!

Another event on the immediate horizon is the launch of Euclid, in 2022. This telescope, operated by the European Space Agency and exclusively devoted to cosmology, will collect data in the spectrum of optical light and the near infrared, at wavelengths between 550 and 900 nanometers. The experiment does not involve our nearby environment so much as a large volume

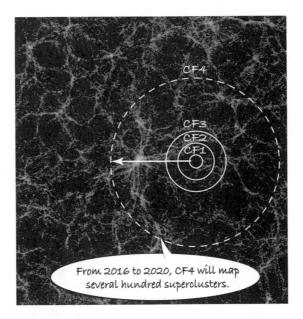

Figure 5.7 The sizes of the four Cosmicflows programs. (Also plate 13.)

much farther away—at distances from 8 to 16 billion light-years—whose long-term evolution it will map. With Euclid, we'll be looking to estimate, among other things, the "growth rate" of large-scale structures. In addition to measuring their size, we'll see them develop. Isn't that incredible? Unlike the Hubble Space Telescope, Euclid will be out of reach once it's in orbit; it's headed for L2, the second Lagrange point, which is located 1.5 million kilometers from Earth, in the direction opposite the sun. From here, the satellite will orbit the sun at exactly the same angular velocity as the earth, in such a way that our planet will constantly protect it from solar interference. Euclid will survey an area covering a third of the celestial sphere. We'll get measurements for 50 million galaxies!

Today, 14 European countries are actively contributing to the preparation and construction of the Euclid satellite (Germany, Austria, Belgium, Denmark, Spain, Finland, France, Italy, Norway, the Netherlands, Portugal, Romania, the United Kingdom, and Switzerland). A total of 1,200 people are working together, including over 750 researchers specialized in astrophysics, cosmology, theoretical physics, and particle physics at over 120 laboratories. Needless to say, the program's success depends on coordinating the work done by researchers, engineers, technicians, and administrative staff. Since the project officially began in 2011, participants have met once a year to discuss the progress in each member country.

At the time of writing, our colleagues in physics and engineering are building the spectroscope, imager, and sensors for the telescope as a whole. Meanwhile, astrophysicists such as myself are optimizing equations and algorithms for measuring the growth rate of galaxy structures. It's a little bit like making a height chart for the size of superclusters at different ages, to see

Figure 5.8 Photos and artist's impression of the telescopes of the future. Top to bottom: ASKAP (Australia), MeerKAT (South Africa), Euclid (European Space Agency).

how tall the "kids" are getting. Knowing these structures' development will also provide information about underlying physical processes. This time, with Euclid, the volume "mapped" will represent about 1% of the observable universe. That's quite a feat! (Remember, by way of contrast, that to date our Cosmicflows research program has mapped only a millionth of the observable universe's volume.) Taking Laniakea's size for reference, we'll map something on the order of several dozens of thousands of superclusters and just as many void regions. In other words, we'll get to know extragalactic structures in a wide array of shapes and sizes: filaments, walls, superclusters, and gravitational watersheds—all of which will aid us in better understanding the distribution of matter and the nature of the forces at play throughout the universe.

Epilogue

We're nearing the end of our extragalactic voyage. I hope you now have a clearer picture of your "local" environment. Even if it's impossible to conceive of the incredible distances at issue. Even if you're not convinced that elusive dark matter, toward which all galaxies are flowing, really exists. Even if it's easy to forget that such falling motion is nothing but a minor irregularity when compared to the vertiginous velocity at which the universe is expanding thanks to dark energy. I can imagine that reading the book at hand has raised many questions without always providing an answer. But if that's so, then I've reached my goal. It's only taken a few hours to encounter all these new ideas. As for me, I've been mapping large-scale structures for over twenty years (already), and I plan to keep doing so for some time to come. This search for the Great Attractor has only led me to marvel more and more at the universe's beauty and intricacy.

Recently, a young high schooler I was telling about my profession asked if it takes a special sensibility to become an astrophysicist. Indeed, I believe one should cultivate a responsive attitude when encountering new information: a capacity to take it in and analyze it—and sometimes, even, to react emotionally. Science isn't solemn. It's so satisfying when an article is accepted

Figure 6.1 From left to right: Daniel Pomarède, Yehuda Hoffman, Hélène Courtois, Brent Tully, Stefan Gottlöber.

for publication! And even though it's frustrating to confront a difficult problem, that doesn't mean we're unhappy: it's another challenge, and we're ready to face it!

Astronomers aren't gentle dreamers who scorn real-world politics and economy. We know exactly how much each sensor, spectroscope, or telescope mirror costs; after all, we need to find the resources it takes to build our instruments. This funding is almost always international. By the same token, the nationality of our colleagues doesn't matter. What counts are complementary scientific skills. Fundamental research transcends national borders, which are artificial.

Figure 6.2 Family photo!

As you've just seen, the discovery of Laniakea was a real group effort. From my very first days as an astrophysicist in Lyon, I've had the good fortune of learning how to gather data from all over the world and then share it with even more researchers. In the process, one member at a time, I've assembled the Cosmic-flows project team, which is exemplified by four researchers with different but complementary skills: Brent Tully (standardizing data), Yehuda Hoffman (theoretical analysis for calculating velocity and density fields), Daniel Pomarède (visualizing results), and myself (observation, data reduction, and analysis). But it's important to remember all the others who have worked, or currently work, on this program. Indeed, Cosmicflows and CLUES—the numerical simulation project directed by Stefan

Gottlöber and Gustavo Yepes—comprise a small family of about forty people across the globe.

The mission of researchers, beyond acquiring new knowledge, is to pass it on to the rest of society. This is why, in addition to conducting research, I teach university courses, give lectures open to the public, and meet with young people. Together with Michel Tognini, who's an astronaut, I'm one of the "godparents" of the Vaulx-en-Velin Planetarium. At this facility, which more than 90,000 people visit each year, we try to pass on the latest astronomical research—to inform, explain, and inspire. It's impossible to overstate the importance of education for instilling the values of liberty, equality, and fraternity. Plus, the more one knows, the better life is! It can't be said enough, education is key for forming free, equal, and caring minds. And most of all, the more we learn, the happier we are! Sailors on the cosmic flows, we examine our new maps of the universe—and then I, for one, am off on another observation campaign, to work at telescopes all across the globe.

I invite you to share and savor the dizzying excitement that research in all fields provides: a sense of wonder at the elegance and complexity of our universe, which drives the human quest for knowledge ever onward...

Acknowledgments

Thank you to Denis and my family for their constant support, enthusiasm, and dedication.

My thanks for the help provided by Anne Bourguignon, Clémence Mocquet, Sarah Forveille, Éric Guégen, and Caroline Bee at Éditions Dunod, and to Jermey Matthews and Matthew Abbate at the MIT Press.

Bibliography/Webography

Videos Linked in This Book

Video 1 (see figure 3.8): https://youtu.be/q10sOK-VBtw; animated figure
   (see figure 3.9): https://skfb.ly/R9pN
Video 2 (see figure 4.6): https://youtu.be/R5BkZsxBwRs
Video 3 (see figure 4.7): https://youtu.be/ThfbMsj8awM
Video 4 (see figure 4.8): https://youtu.be/E1eziUa0iFI
Video 5 (see page [73]): https://youtu.be/rENyyRwxpHo
Video 6 (see figure 4.13): https://youtu.be/fdBvuNMBBtI
Video 7 (see figure 5.2): https://youtu.be/HS1xzsOOENY
Video 8 (see page [90]): https://youtu.be/W0Bd5Vx0Uls

Books

Beaudoin, Emmanuel. *101 merveilles du ciel qu'il faut avoir vues dans sa vie*. Paris: Dunod, 2015.

Bernardeau, Francis. *Cosmologie: Des fondements théoriques aux observations*. Paris: CNRS éditions, EDP Sciences, 2007.

Bertone, Gianfranco. *Le mystère de la matière noire*. Paris: Dunod, 2014.

Combes, Françoise. *La matière noire, clé de l'Univers?* Paris: Vuibert, 2015.

Combes, Françoise. *Mystères de la formation des galaxies. Vers une nouvelle physique?* Paris: Dunod, 2008.

Combes, Françoise, and James Lequeux. *La Voie lactée*. Paris: CNRS Éditions, 2013.

Gott, J. Richard. *The Cosmic Web: The Mysterious Architecture of the Universe*. Princeton, NJ: Princeton University Press, 2016.

Haddad, Leïla, and Guillaume Duprat. *Mondes. Mythes et images de l'univers*. Paris: Seuil, 2006.

Lachièze-Rey, Marc. *Initiation à la cosmologie*. 6th ed. Paris: Dunod, 2016.

Robert-Esil, Jean-Luc, and Jacques Paul. *Le petit livre de l'Univers*. Paris: Dunod, 2014.

Schüler, Chris. *La mer et les étoiles. La cartographie maritime et céleste de l'antiquité à nos jours*. Paris: Place des Victoires, 2012.

Research Articles in English

Courtois, Hélène, et al. "Cosmicflows-3: Cold Spot Repeller?" *Astrophysical Journal* (2017): http://arxiv.org/pdf/arXiv:1708.07547.pdf

Courtois, Hélène, et al. "Cosmography of the Local Universe." *Astronomical Journal* 146, 69 (2013): http://iopscience.iop.org/article/10.1088/0004-6256/146/3/69/pdf

Hoffman, Yehuda, et al. "Cosmic bulk flow and the local motion from cosmicflows-2." *Monthly Notices of the Royal Astronomical Society* 449, 4494 (2015): https://arxiv.org/pdf/1503.05422v1.pdf

Hoffman, Yehuda, et al. "The Dipole Repeller." *Nature Astronomy*, 1, 36 (2017): https://arxiv.org/pdf/1702.02483.pdf

Hubble, Edwin. "A relation between distance and radial velocity among extra-galactic nebulae." *Proceedings of the National Academy of Sciences of the USA* 15(3):168–173

Pomarède, Daniel, et al. "The Arrowhead mini-supercluster of galaxies." *Astrophysical Journal* 212, 17 (2015): http://arxiv.org/pdf/1509.02622v1.pdf

Pomarède, Daniel, et al. "The Cosmic V-Web." *Astrophysical Journal* 845, 55 (2017): http://arXiv.org/pdf/1706.03413.pdf

Tully, Brent, et al. "The Laniakea supercluster of galaxies." *Nature* 513, 7516 (2014), 71: https://arxiv.org/ftp/arxiv/papers/1409/1409.0880.pdf

Extragalactic Databases

EDD, Extragalactic Distance Database: http://edd.ifa.hawaii.edu/

LEDA, Lyon Extragalactic DAtabase (or HYPERLEDA): http://leda.univ-lyon1.fr/

NED, NASA IPAC Extragalactic Database: http://ned.ipac.caltech.edu/

Further Videos

The coldest point in the cosmic microwave background, interactive figure: https://skfb.ly/6sXCY

Constrained numerical simulations: https://www.clues-project.org, http://irfu.cea.fr/localuniversesimulation

The cosmic velocity web: http://vimeo.com/pomarede/vweb (interactive version: https://skfb.ly/667Jr)

Cosmography with CF1 with 3D stereo visualizations: http://irfu.cea.fr/cosmography (2013)

Cosmography with CF2: http://irfu.cea.fr/laniakea (2014), http://irfu.cea.fr/arrowhead (2015)

The Dipole Repeller: http://irfu.cea.fr/dipolerepeller

TEDx Lyon conference: https://youtu.be/W4jIApDM3so (2016)

Vaulx-en-Velin planetarium: https://youtu.be/J5ID_7BGSSI (2014), https://youtu.be/FJpJ0sNUKrk (in German)

Image Credits

1.1 Left: NASA, ESA, and The Hubble Heritage Team (STScI/AURA); right: NASA, ESA, and The Hubble Heritage Team (STScI/AURA); bottom: ESO.

1.2 Top left: NASA/JPL; top right: ESO; bottom left: Springel et al. (2005); bottom right: Hélène Courtois and Benjamin Le Talour.

1.8 Harvard College Observatory/Science Photo Library.

1.11 Photo by René Cavaroz.

2.8 Courtesy of University of California, Santa Cruz.

2.11 Solaris.

3.2 Photo by John Zich.

3.3 Top left: photo by John Sarkissian (© CSIRO Parkes Observatory); top right: NRAO/AUI/NSF; middle left: photo by Richard Wainscoat © Institute for Astronomy; middle right: NASA/JPL-Caltech/R. Hurt (SSC); bottom left: STS-8 Crew, STScI, NASA; bottom right: Courtesy of the NAIC—Arecibo Observatory, a facility of the NSF.

3.6 ESO.

4.4 Top: DMR, COBE, NASA, Four-Year Sky Map; bottom: ESA and the Planck Collaboration.

4.5 Courtesy of Carnegie Institution of Washington.

4.12 Photo by Michel Di Nella.

5.2 Courtesy of Jenny G. Sorce.

5.6 Photo by Sean Wilson.

5.8 Top: CSIRO; middle: Picture Alliance/dpa; bottom: ESA/C. Carreau.

6.1 Photo by Noam Libeskind.

6.2 Photo by Roland Triay.

Index